Dedication

To my mother Irene, who kept me fed in England during the food-rationed years following World War II, and at whose knee I had my first lessons in nutrition—and to my father Rufus who earned the money to make it possible.

Contents

About the Author

John Ashley is a practitioner of food security at the levels of formulating national food security strategies, project design and field implementation, and evaluating programs and projects which have sought to increase resilience to food insecurity. He has also engaged with agricultural research and University teaching. He graduated in botany from London University, and then applied that basic training to the field of agriculture for his doctorate, working with the groundnut crop at Makerere University, Uganda. He also holds a degree in psychology from Cambridge. He has worked in some 20 vulnerable and/or conflict-prone countries over 40 years. He married into Uganda and, when not on consulting missions in other countries, helps his wife and other family members manage their 65-acre farm there, which supports crop, livestock and forest components.

Ensuring National Biosecurity

Ensuring National Biosecurity

Institutional Biosafety Committees

Edited By

Carole R. Baskin

Alan P. Zelicoff

Institute for Biosecurity,
College for Public Health and Social Justice,
Saint Louis University,
St Louis, MO, USA

AMSTERDAM • BOSTON • HEIDELBERG • LONDON
NEW YORK • OXFORD • PARIS • SAN DIEGO
SAN FRANCISCO • SINGAPORE • SYDNEY • TOKYO
Academic Press is an imprint of Elsevier

Academic Press is an imprint of Elsevier
125, London Wall, EC2Y 5AS.
525 B Street, Suite 1800, San Diego, CA 92101-4495, USA
225 Wyman Street, Waltham, MA 02451, USA
The Boulevard, Langford Lane, Kidlington, Oxford OX5 1GB, UK

Notices
Knowledge and best practice in this field are constantly changing. As new research and
experience broaden our understanding, changes in research methods, professional practices,
or medical treatment may become necessary.

Practitioners and researchers must always rely on their own experience and knowledge in
evaluating and using any information, methods, compounds, or experiments described herein.
In using such information or methods they should be mindful of their own safety and the
safety of others, including parties for whom they have a professional responsibility.

To the fullest extent of the law, neither the Publisher nor the authors, contributors, or editors,
assume any liability for any injury and/or damage to persons or property as a matter of
products liability, negligence or otherwise, or from any use or operation of any methods,
products, instructions, or ideas contained in the material herein.

ISBN: 978-0-12-801885-9

British Library Cataloguing-in-Publication Data
A catalogue record for this book is available from the British Library.

Library of Congress Cataloging-in-Publication Data
A catalog record for this book is available from the Library of Congress.

For Information on all Academic Press publications
visit our website at http://store.elsevier.com/

Typeset by MPS Limited, Chennai, India
www.adi-mps.com

  Working together
to grow libraries in
developing countries

www.elsevier.com • www.bookaid.org

Publisher: Mica Haley
Acquisition Editor: Erin Hill-Parks
Editorial Project Manager: Molly McLaughlin
Production Project Manager: Lucía Pérez
Designer: Greg Harris

Contents

List of contributors

Jeffery Adamovicz Laboratory for Infectious Disease Research, University of Missouri, Columbia, MO, USA

Ryan Bayha NIH Office of Biotechnology Activities, Office Science Policy, National Institutes of Health, Bethesda, MD, USA

R. Mark Buller Department of Molecular Microbiology and Immunology, Saint Louis University, St. Louis, MO, USA

Ryan N. Burnette International Biosafety & Biosecurity Programs, AT-RISK International

Mark Campbell Department of Molecular Microbiology and Immunology, Saint Louis University, St. Louis, MO, USA

Nick Chaplinski United States National Poultry Research Center, Athens, GA, USA

Nancy D. Connell Department of Medicine, Rutgers New Jersey Medical School, Newark, NJ, USA

Susan Cook Washington University, St. Louis, MO, USA

Jacqueline Corrigan-Curay NIH Office of Biotechnology Activities, Office Science Policy, National Institutes of Health, Bethesda, MD, USA

Kathryn L. Harris NIH Office of Biotechnology Activities, Office Science Policy, National Institutes of Health, Bethesda, MD, USA

Deborah Howard Regulatory Affairs Seeds & Traits, Bayer CropScience LP, Efland, NC, USA

Jens H. Kuhn Tunnell Government Services, NIH/NIAID Integrated Research Facility at Fort Detrick, Frederick, MD, USA

Charles E. Lewis USDA National Veterinary Services Laboratories, Ames, IA, USA

M. Malendia Maccree Agriculture and Natural Resources, University of California, Davis, Davis, CA, USA

William S. Mellon Emeritus Professor of Pharmaceutical Sciences, Former Associate Dean for Research Policy, Graduate School, University of Wisconsin-Madison, Madison, WI, USA

Steven S. Morse Department of Epidemiology, Mailman School of Public Health, Columbia University, New York, NY, USA

David Rainer Environmental Health and Public Safety, North Carolina State University, Raleigh, NC, USA

Raymond C. Tait Department of Neurology and Psychiatry, Saint Louis University, St. Louis MO, USA

David M. White National Centers for Animal Health, Ames, IA, USA

Bruce Whitney Texas A&M University, College Station, TX, USA

Carrie D. Wolinetz NIH Office of Biotechnology Activities, Office Science Policy, National Institutes of Health, Bethesda, MD, USA

Alan P. Zelicoff Institute for Biosecurity, College for Public Health and Social Justice, Saint Louis University, Saint Louis, MO, USA

Preface

In recent months, the White House Office of Science and Technology Policy identified emerging technologies, global health security, and homeland security and intelligence as critical components of science and technology investment in the United States. Although these investments involve many scientific sectors and disciplines, one underlying principle applies to all: the need to promote research that enhances national interests and initiatives while balancing the costs of conducting it.

As we are writing this, the US Government is orchestrating an intense public discourse over contentious matters in the biosecurity community, including policies governing Dual Use Research of Concern and the Select Agent Regulations, thus demanding careful assessment of the impacts, both positive and negative, of science policy on US biosecurity, research enterprise, and society. The tradeoffs are complex and dynamic, consistent with the nature of scientific inquiry and, not to be forgotten, the political landscape as well.

When an acquisition editor at Elsevier approached us to write a book on national biosecurity and the increasing (and mostly unfunded) responsibilities and workload of Institutional Biosafety Committees (IBCs), we had serious doubts that we would be able to find a sufficient cohort of experts interested in contributing. After all, given the topic of the book, professionals of this ilk are already terribly overcommitted. Yet when we started making requests, the response was overwhelmingly positive: it turned out that biosafety practitioners, high-level administrators, and scientists had much to say on this topic, and were eager to contribute. As members of a teaching and research faculty, we heard plenty of cautionary tales from colleagues about the challenges of editing a multi-authored book, but given rapid evolution of the scope of responsibilities of IBCs, we felt that this volume might serve as a timely resource for policymakers and researchers alike.

To the contributors to this book, we know well that you have gone to great lengths to write and revise chapters, work that added to your daily professional responsibilities while forsaking time with family and friends. Because of your efforts, we believe that this compendium will contribute meaningfully to the ongoing conversations about laboratory biosafety and biosecurity. Thank you.

Carole R. Baskin and Alan P. Zelicoff

Laboratory biosecurity in the United States: evolution and regulation

1

Alan P. Zelicoff

Chapter Outline

Introduction

The term "biosecurity" is used in a wide variety of contexts and carries with it an equally diverse set of meanings. For example, veterinarians traditionally view biosecurity as the set of management practices to protect animals – livestock or others of economic value – against microbial threat, some of which may be inadvertently introduced by humans. Preventing influenza in pig farming and tuberculosis (*Mycobacterium tuberculosis*) among elephants in zoological parks are two illustrations [1]. "Biosecurity" takes on an entirely different meaning in international political agreements such as the Biological and Toxin Weapons Convention of 1975, where it refers to measures to prevent the research and development of microorganisms or their products for hostile purposes [2]. And it is not too far a reach to think of biosecurity as the prevention of infectious disease – and specifically communicable infectious disease – in humans [3].

Ensuring National Biosecurity. DOI: http://dx.doi.org/10.1016/B978-0-12-801885-9.00001-9

For the purposes of this chapter and those that comprise the balance of this text we will employ the definition promulgated by the US Department of Health and Human Services [4]:

> *[T]he term biosecurity refers to the protection, control of, and accountability for high-consequence biological agents and toxins and critical relevant biological materials and information within laboratories to prevent unauthorized possession, loss, theft, misuse, diversion, or intentional release. Biosecurity is achieved through an aggregate of practices including the education and training of laboratory personnel, security risk assessments, Biological Select Agent and Toxin (BSAT) access controls, physical security (facility) safeguards, and the regulated transport of BSAT.[1] Achieving effective comprehensive biosecurity for BSAT is a shared responsibility between the Federal Government and facilities/individuals that possess, use or transfer BSAT.*

Complementary to, but distinct from, biosecurity is *biosafety* based on principles of *containment* and *risk assessment* in the laboratory. Containment includes: "the microbiological practices, safety equipment, and facility safeguards that protect laboratory workers, the environment and the public from exposure to infectious microorganisms that are handled and stored in the laboratory," whereas risk assessment is "the process that enables the appropriate selection of microbiological practices, safety equipment, and facility safeguards that can prevent laboratory-associated infections" [5].

A helpful means of distinguishing "biosecurity" and "biosafety" is to note that they commonly differ on *intent*, that is, biosecurity is implemented to obviate the intentional diversion or release of biological materials, whereas biosafety measures limit their unintentional dissemination in order to protect laboratory workers and the surrounding community and environment from accidental exposure to pathogens [6,7]. The functional components of biosecurity architecture will be described below.

The purpose of this chapter is to review the evolution of biosecurity and modern tenets of its implementation as it applies to high-containment laboratories or those working with "select agents" as defined by statute. Many (if not most) laboratorians are unaware of the historical origins of biosecurity. Perhaps of greater importance is that laboratory officials and researchers working with dangerous pathogens may be naïve to the origins in the law of the now lengthy list of operational biosecurity requirements, obviously of practical relevance in the day-to-day functions of research facilities. The key pieces of legislation that have mandated these requirements were responses to events such as bioterrorism threats in the late 1990s and the downing of the World Trade Center buildings in 2001 as we shall see in more detail shortly.

Biosecurity laws passed by the Congress vest considerable authority in government departments such as Health and Human Services (HHS) and Agriculture (USDA) to formulate and then implement regulations (frequently referred to by officials as "rules") with which laboratory workers, researchers, staff and security personnel must comply. These rules are revised at intervals, sometimes on a regular basis and also

---

[1] Select agents and toxins will be described in detail below.

when new laws are passed. We will summarize the processes by which agencies with HHS and USDA – typically the Centers for Disease Control and Prevention (CDC) and the Animal and Plant Health Inspection Service (APHIS) – interpret the will of Congress (via laws that have been proposed, debated and passed), formulate proposed regulations, solicit comments from individuals and entities likely to be affected, and then disseminate final rules. Beyond the legislation itself, the president may issue directives: these include executive orders or "EO"s, which have the full force of the law and must be published in the Federal Register (FR); and administrative orders such as memorandums, determinations, notices, which have the same legal effect but do not have a publication requirement in the FR and may therefore be "born classified." All of these may prompt executive agencies (such as HHS and USDA) to craft new rules as well. Several biosecurity-relevant EOs will also be reviewed.

But neither Congress nor the Executive Office of the President act without also taking into account the advice – sometimes directly solicited, sometimes not – of subject matter experts in academia and professional practice. Thus, over the past few decades there have also been several key reports from professional organizations, *ad hoc* groups and government-sponsored panels that have had a dramatic influence on biosecurity practice. Their importance goes beyond mere operational standards for laboratories. Rather, documents such as those produced at the ground-breaking Asilomar Conference in 1975 [8] through the recent publications of National Science Advisory Board for Biosecurity (NSABB) and the Federal Experts Security Advisory Panel have set in motion an inclusive process for scientists in and outside of government to recommend revision of biosecurity requirements that reflect research priorities involving naturally occurring organisms and (regrettably) those which might be used in bioterrorism. Because of their importance to laboratorians these reports will also be summarized.

In the end, the detailed regulations now extant in laboratories where certain pathogens and toxins – those dangerous to humans, animals or plants if released either inadvertently or intentionally – are kept for research purposes came about as a result of the complex interaction of public apprehension expressed in Congressional legislation and EOs, technical analysis from scientists and expert groups, and practical concerns from researchers who seek to carry out noble work in disease prevention and treatment. The now famous "Select Agent and Toxin List" (SATL) is perhaps the most visible result of the regulatory framework that applies to many biological laboratories and we will show its development over the past decade-and-a-half in detail. In so doing, we hope to foster involvement of thoughtful scientists in formulating policy. After all, bench scientists often have far more familiarity with cutting-edge research and experience with laboratory practices than most officials in the executive branch of government tasked with enacting far-reaching legislation.

Finally, since the turn of the century there have been a few high-profile "near misses" where the breakdown of biosecurity in containment laboratories could have resulted in infections among personnel or the public. Investigations directed at root-cause analysis often result in additional regulatory restrictions with both direct and indirect costs. We will attempt to weigh their benefits against perceived and real costs.

Historical origins of biosecurity

In the mid-nineteenth century a series of "International Sanitary Conferences" were held in Paris, Vienna, Constantinople, Washington, Rome and Dresden with the goal of interrupting recurring epidemics of three diseases recently arrived or reappearing in Europe and North America: cholera, plague and yellow fever (each disease entity had a clinical pedigree and its epidemiologic characteristics roughly described, though all were without known cause). Over the course of 14 meetings starting in 1851 and ending in 1938, participants from the medical and diplomatic communities debated the origins of these diseases and the preventive actions that could be taken to "protect [people] and control" biological agents. This was the naissance of biosecurity in its most straightforward sense, and in retrospect is surprising given that the germ theory of disease was, at that moment in history, barely being formulated and understood. Absent that theory, early on in the Sanity Conferences, physicians and diplomats representing about a dozen countries from the United States to Russia argued over the effectiveness of quarantine and the very nature of what is now recognized as infectious disease. Anthony Perrier of Great Britain declared at the first gathering that cholera was not "communicable" and that "contagion is not a fact, but a hypothesis invented to explain a number of facts that without this hypothesis would be inexplicable" [9]. Offering no better explanation himself, Perrier went on to note that his colleagues "persisted in the routine path of practices that are outmoded, useless and ruinous to commerce and harmful to public health in that instead of enlightening the peoples on the true means of guaranteeing themselves against epidemics, they inspire on the contrary a false sense of security that prevents them from taking the only sanitary precautions that can offer real guarantees." Perrier did not at this juncture specify what might constitute those "guarantees."

This is perhaps the first association of the words "security" and "public health" in the setting of (then unknown) infectious diseases. Remarkably enough, less than 80 years after Perrier's confusing admonitions, the origin and routes of transmission for all three diseases had been identified and effective preventive practices put into routine use – "biosecurity" by any other name. As the Sanitary Conferences continued to meet, in December 1907 representatives decided to formalize the forum into the "Office International d'Hygiène Publique" (OIHP), ultimately subsumed into the League of Nations at the end of World War I. It became known as the League's "Health Organization" and produced an astonishing body of work including outbreak control and mitigation (with locales that ranged from Europe to ports in the Far East), nutrition (across the age spectrum from infants to adults), standards for medications, vitamins, antitoxins and vaccines, epidemiology of cancer (of a variety of organs) and even building construction guidelines to name but a small portion of their work [10]. The OIHP continued to operate until 1946 when the nascent World Health Organization (WHO), today's premier international health security institution, subsumed its functions [11].

Though more than seven decades would pass from the inception of the OHIP, in 1983 the WHO published the Laboratory Biosafety Manual establishing standards for worker safety and laboratory practices. By the time of the third edition in 2004 the Manual evolved to include succinct definitions of biosafety and biosecurity. "'Laboratory biosafety' is the term used to describe the containment principles, technologies, and practices that are implemented to prevent unintentional exposure to pathogens and toxins or their accidental release. 'Laboratory biosecurity' refers to institutional and personal security measures designed to prevent the loss, theft, mis-use, diversion or intentional release of pathogens and toxins" [12]. The WHO Manual further describes biocontainment (including biocontainment levels) and risk assess-ment as the foundations of biosecurity, which in turn have informed US biosecurity strategies in legislation and laboratory practice.

Elements of modern laboratory biosecurity

In its common application and also as enshrined in various US laws, biosecurity is generally taken to be comprised of five or six main components (depending somewhat on definitions), all designed to limit access to pathogens and toxins to prevent their loss, theft or misuse [5,13]:

- Risk assessment that is a detailed listing of the hazardous characteristics of an organism or toxin, the probable consequences of unwanted exposure and associated occupational health plans.
- Access control equipment and barriers, perhaps including perimeter and internal monitoring.
- Personnel reliability, which may include background investigations, medical screening and assessment of expertise and experience.
- Control and accountability of materials (with associated documentation of archived materials).
- Training and emergency planning.
- Program management and supervision.

As noted earlier, none of these constituents is completely unique to biosecurity. Successful biosafety programs also depend to one extent or another on the same pro-cesses and physical constructs.

Key legislation in biosecurity (or how Congressional intent defines biosecurity)

While most laboratory managers and scientists working with pathogens and tox-ins are aware of the numerous regulations that govern access to and use of those materials, fewer understand the legal processes by which these come into effect. It is useful to understand the source of legislative action that lead directly to many of the current laboratory biosecurity/biosafety rules and procedures because scientists have the opportunity to influence the rule-making process even (and some would say "especially") after US Congress passes new laws. We begin with a brief review of the legislative process starting with its inception in Congress through the rule-making procedures that actually implement the law. With this knowledge in mind, we can then put into context the origin of the SATL, now familiar even to those infectious disease researchers who do not necessarily work with these organisms and toxins.

Legislation and its implementation through "rule-making"

The Constitution of the United States vests "all legislative Powers" in "a Congress of the United States".[2] Any member of Congress may introduce legislation, and such proposals are usually referred to as "bills." Bills originating in the House are designated as "House Resolutions," and carry the abbreviation "HR" before the unique number assigned to it. Similarly in the Senate one finds "S," that is "Senate" (with "Resolution" omitted) for bills proposed by one or more its members. Thus, when tracking the course of a bill through Congress, it is convenient to specify the "HR" or "S" number.[3] Both HR and S require approval by the other body and the signature of the president to become law.

After a bill is introduced, Congressional committees almost always hold formal hearings designed to gather information about the impact of the bill on already existing laws and any new requirements it imposes, and costs (if money has to be appropriated to fund the bill). Committee chairmen invite both private and government experts to testify (especially from departments or agencies that will help write the regulations that implement the bill if it passes), and in the process the bill is typically changed (or "marked up") before the committee takes a vote to either move the bill forward or not. It is not unusual for a full committee to refer a bill for discussion to a subcommittee.

Should a bill be "reported out of Committee" – meaning it is referred to the full body of the House or Senate – it is then debated "on the floor" where any member can request time to speak. Generally, after debate has completed, amendments can be offered, and then the bill is voted upon. A similar version of the bill goes through the same process in the other body, and if passed by a majority of both houses in the same form, is then sent to the president for signature.

There are, of course, complications that frequently derail the more-or-less straightforward description above, often deriving from the complex committee structure in the Congress. There are currently 26 committees in the House[4] and 24 in the Senate.[5] With rare exception, each committee has several subcommittees (as of this writing a total of 94 subcommittees in the House and about 70 in the Senate), and subcommittees often have overlapping jurisdiction, so several may hold hearings on a given bill simultaneously. Thus, only a minority (around 15%) of introduced bills are reported out of committee for vote on the floor of House and Senate, and few bills escape the

---

[2] Constitution of the United States, Article 1 Section 1.

[3] One website for following the progress of bills through introduction, committee deliberations, floor debate and voting is: https://www.govtrack.us/congress/bills/

[4] http://www.house.gov/committees/

[5] http://www.senate.gov/pagelayout/committees/d_three_sections_with_teasers/committees_home.htm

committee process without substantial rewriting of the original proposed text based on the decisions of committee members after listening to witnesses at hearings, or taking into account the views of their constituents. And even after surviving this process, the senior leadership of the House and Senate each decide whether or not they will, in fact, allow a bill to come up for floor debate at all. Should a bill pass both houses in something other than identical form – as often occurs when one house makes amendments to a bill originating in the other body – a "conference committee" must reconcile the differences. If (as is not unusual) a dozen or more subcommittees from each house have been involved in discussing the bill, more than a hundred members may appear at the conference committee (or "conference" for short) meetings.

After the conference negotiates the differences in House and Senate versions of the Bill, it is sent back to both bodies for a final vote; usually the bill is passed after all effort described above is completed. Before printing of the bill, it receives a numerical designation of the session of Congress and the number of the bill for that session. For example, the "Antiterrorism and Effective Death Penalty Act of 1996" (the "popular name" of the original bill) was originally introduced in the Senate of the 104th Congress and (in session from 1995 to 1996) as S735, and was the 132nd piece of legislation considered, so when passed it received the numerical indicator 104-132, in addition to its popular name, which is then abbreviated as "Public Law 104-132" or "PL 104-132".

The president may sign the bill or choose to veto it, and Congress may override that veto with a two-thirds vote in both houses. Upon the president's signature, or in the case of a veto followed by a Congressional override, a "bill" becomes a "law" and the various provisions of the law "statutes." The vast majority of laws are denoted as "Public Laws" which means that they apply to individuals and their relationship with government or society. (There are also "Private Laws" applying to the relationships between individuals, such as contracts).

Laws are, in essence, codes of conduct, and Public Laws (also called "Acts") often impose new rules for behavior of individuals, companies or institutions. Public Laws almost always also make modifications to existing laws that comprise the United States Code (formally abbreviated as "USC"), currently arranged in 51 "Titles," really sections of law covering familiar aspects of life such as commerce and trade, crimes and criminal procedure, copyrights, food and drugs, taxes, foreign relations, alcohol and firearms, banks and transportation to name but a few.[6] Of particular interest in biosecurity law is Title 42 – "The Public Health and Welfare" also denoted as "42 USC" – that is comprised of many hundreds of sections. So, perhaps unsurprisingly PL 104-132 made more than a dozen changes to Title 42 since laboratory safety and security naturally impact the public's health. But it also made changes to 18 USC ("Crimes and Criminal Procedure").

New Public Laws routinely mandate actions to be taken by cabinet departments in the executive branch of government, such as HHS. It is then the responsibility of the cabinet secretary to implement those actions. As we will see shortly, with recent biosecurity-related legislation, the secretaries of both HHS and USDA are now

[6] http://www.law.cornell.edu/uscode/text

required to formulate and update a list of organisms and toxins that may be of particular importance to public health if inadvertently released or misused (for example
in a biological weapon).

How does this implementation happen in practice? In order to execute new laws,
the Secretary (one or more are always specified in the law) designates an agency
within her department to publish an initial proposal indicating the intent of the executive branch to carry out the will of Congress, and it appears in the FR as a "notice of
proposed rule-making" (NPRM), often within days of the president's signature on the
original Act. It is worth noting that the Secretary is granted latitude in interpreting
Congressional intent, and as we shall see exercises considerable judgment in publishing the NPRM.

The FR is closely read by administrators in business, government, law, and law
enforcement, along with individuals who may be affected by the new PL.[7] Via the
announcement in the FR, any interested party may submit comments or critiques of
the agency's proposed means of implementing the law. Comment periods typically
last 30–60 days (and may even be supplemented by public meetings if the new regulations that result from the rule-making procedure are sweeping enough), after which
the agency assembles the suggestions and testimonies. A "Final Rule" is published in
the FR as soon as the agency adjudicates the (frequently disparate) collection of
views, which then becomes new regulation. The agency and cabinet secretary are
under no obligation to accept any particular individual or individual entity's views,
and by no means is the decision on the structure and requirements of the new regulation a matter of simply tallying the net opinion of commenters. Rather, the agency
uses its own experts – including attorneys who interpret the oft-subtle intention of the
Congress – in formulating the regulation. It may take many months for the cabinet
department to publish the Final Rule.

Shortly after the Final Rule appears in the FR, the new regulation is enshrined in the
Code of Federal Regulations (the CFR, not to be confused with the United States Code)
where it remains in force until a review is ordered by Congress, or if a new legislation
includes provisions for updating the regulation; agencies may also publish proposed
updates in the FR and solicit additional comments from individuals or organizations
likely to be affected. The CFR is, like the USC, organized by "Title" whose names *mostly*
parallel the Titles in the US Code (unfortunately, this is not always the case).

In summary, after a bill is introduced into Congress, debated and then enacted
into law by signature of the president, executive agencies (e.g. HHS or Department
of Agriculture) are then mandated to implement the detailed requirements by a "rule-
making" process that is initiated with a NPRM published in the FR. Individuals,
organizations, businesses or other entities are invited to comment on the agency's
initial plans. These comments are considered by the agency tasked with crafting the
new rule, and then published as a "Final Rule" that is, in practice, the set of regulations that carry out the will of Congress which are then recorded in the CFR.

---

[7]Announcements in the Federal Register are designated by volume number and the first page of the
announcement. For example, "61 FR 29327" starts at page 29,237 of the 61st volume of the Federal
Register. In general, any announcement may be found online at: https://www.federalregister.gov/

Historical origins of current biosecurity regulations

On April 19, 1995 the Alfred P. Murrah Federal Building in Oklahoma City was destroyed by a truck bomb planted by Timothy McVeigh, a former solider and militia movement sympathizer. This event, which resulted in 169 deaths and hundreds of injuries and property damage in excess of a half-billion dollars, stood as the deadliest terrorist attack on US soil until the downing of the World Trade Center 6 years later.

With the memory of yet another terrorist attack – the 1993 World Trade Center bombing – still fresh, President Bill Clinton had already introduced antiterrorism legislation in early 1995 (called "the Omnibus Counterterrorism Act"), but within days of the Oklahoma City event Senate Majority Leader Robert Dole was motivated to propose a similar but more sweeping bill, "The Comprehensive Terrorism Prevention Act," S735 [14].[8] When initially introduced, the most prominent component of the Act was a limitation on habeas corpus actions brought to federal court by prisoners suspected of an act of terror. When finally passed by the Congress as The Antiterrorism and Effective Death Penalty Act (ATEDPA) of 1996 exactly 1 year to the day after the Oklahoma tragedy, the new law (PL 104-132) included requirements for the Secretary of HSS to:

- "Establish and maintain a list of each biological agent that has the potential to pose a severe threat to public health and safety," based on specific criteria including effect on human health, degree of contagiousness, availability and effectiveness of immunization and treatments for illness caused by the agent, and "any other criteria that the Secretary considers appropriate in consultation with scientific experts."
- Regulate transfers of listed biological agents including establishing and enforcing safety procedures, safeguards to prevent access to listed agents for use in terrorism or other criminal purposes while maintaining "appropriate availability of biological agents for research, education, and other legitimate purposes" [16].

The Act was signed into law by President Clinton on April 27, 1996. On June 10th of the same year, the CDC, acting on behalf of the Secretary of HHS published in the FR a "NPRM" soliciting comment on implementing the new requirements of PL104-132 to ensure public safety, strengthen public–private sector accountability and collect information concerning the location of potentially hazardous infectious agents while tracking the acquisition of those agents.[9] The SATL was born. The CDC also

---

[8] By 1996 there were at least two other factors motivating the Congress. In May of that year Larry Wayne Harris a member of Aryan Nation and a self-styled biological weapons expert purchased several vials of *Yesinia pestis*, the causative organism of plague from the American Type Tissue Collection a microbiologic supply house in Maryland. Harris was arrested but at the time there was no US law prohibiting individuals from purchasing disease-causing organisms so officials charged him with mail fraud because he misrepresented himself as the operator of a legitimate medical laboratory. Also a few weeks before Mr Harris' arrest on March 20, 1995 members of Aum Shinrikyo, an apocalyptic cult in Japan, released an unknown quantity of sarin gas in the Tokyo subway system. Thirteen people were killed and more than 6000 individuals with varying symptoms (some of which were surely related to panic alone) were seen in hospital emergency rooms [15].

[9] http://www.gpo.gov/fdsys/pkg/FR-1996-06-10/pdf/96-14707.pdf

proposed procedures for alerting law enforcement of unauthorized attempts to acquire select agents.[10] A 30-day comment period permitted interested parties to address:

- the content of the "Select Infectious Agents List" initially comprised of 26 organisms (viruses, bacteria, *Rickettsiae* species and fungi) and 14 toxins,[11]
- registration *and inspection* of facilities transferring Select Infectious Agents,
- transfer requirements (including transfer forms),
- verification procedures including requiring facilities to have a "Responsible Official" (RO),
- agent disposal requirements,
- exemptions for biosafety level 2 and clinical laboratories.

The final rule took into account over 200 written comments and was published in the FR on October 24, 1996.[12] The SATL (changed from "Select Infectious Agents List") was comprised of 13 viruses or virus groups, seven bacteria species, three *Rickettsiae* species, *Coccidioides immitis* as the sole fungal species and 12 toxins. At the same time, the CDC also informed entities owning select agents that it would provide application forms for facility registration with the possibility of facility inspection depending on documentation provided and agent transfer forms. The Federal Bureau of Investigations (FBI) (and perhaps other federal agencies) would have access to records and databases for law enforcement purposes.

The final rule took effect on April 15, 1997 and was placed in the CFR Title 42 Part 72 (later moved to Part 73, see below).[13]

Thus, the history of creating the initial SATL and its associated reporting and implementation requirements was:

- Introduction of "The Comprehensive Terrorism Prevention Act" to Congress as a result of terrorist attacks on US soil.
- Passage of a markedly revised bill renamed as "The Antiterrorism and Death Penalty Act" of 1996 and signed into law as PL 104-132 by President Clinton.
- Changes are made to the US Code and in particular Title 42 of the US Code, "Public Health and Welfare".[14]
- Because the Secretary of HHS was required by PL 104-132 to regulate transfer of certain dangerous pathogens and toxins, the CDC, an agency within HHS and acting on its behalf, publishes a proposed Rule Making for implementation in the FR.
- The CDC accepts comments from any interested party, since the proposed rule was only that – a proposal – written by technical experts in the department.
- After comments were received the CDC made decisions for implementation of the PL (it did not have to accept any particular suggestion, nor did it accept most) and published a Final Rule. The "Final Rule" included the date when the new regulations will go into effect.

---

[10] http://www.gpo.gov/fdsys/pkg/FR-1996-06-10/pdf/96-14707.pdf
[11] As a starting point, the CDC adopted the list of organisms and toxin on the "Australia list", a pre-existing export control regulation limiting the shipment of potentially dangerous biological materials to only selected states (initially comprised of 15 countries). See http://www.australiagroup.net/en/origins.html
[12] http://www.gpo.gov/fdsys/pkg/FR-1996-10-24/pdf/96-27082.pdf
[13] The entire Code of Federal Regulations is available electronically at: http://www.ecfr.gov/. The most recent version of the Select Agents List may be seen at: http://www.ecfr.gov/cgi-bin/text-idx?SID=08582c fb436f670fc3796e016a114198&node=42:1.0.1.6.61&rgn=div5
[14] http://www.law.cornell.edu/uscode/text/42

- The rule was added to US CFR. In this case Title 42 of the CFR part 73 (abbreviated in legal parlance as "42 CFR 73" now entitled as the "Select Agents and Toxins" section of the CFR) specifically part 73.3.[15]

A very similar series of legislative and regulatory events took place after the terror attacks of September 11, 2001. Less than 2 months after that horrific day, the Congress passed the *Uniting and Strengthening America by Providing Appropriate Tools Required to Intercept and Obstruct Terrorism* [17] (abbreviated as the USA PATRIOT Act in a rather tortured acronym). While not directly impacting laboratory biosecurity, the Act defines possessing a biological weapon as a crime and also defined a "restricted person" as one who may not ship or transport any biological agent or toxin that is listed as a select agent (adding to Title 18 of the US Code "Crimes and Criminal Procedures" an entirely new section in "Biological Weapons" chapter called "Possession by restricted persons").

But 9/11 led to yet another milestone in biosecurity law in the following year. The *Public Health Security and Bioterrorism Preparedness and Response Act of 2002* (PHSBRA) [18] was passed in Congress and signed by the president in June. From the standpoint of biological laboratory management and work the Act required the following:

- The Secretaries of *both* HHS and USDA must undertake a *biennial review* of select agents and are instructed to consider agents that should be added or removed from the SATL. The purpose of including the Department of Agriculture was so that biological materials dangerous to animals or plants could be included in the SATL. This would thus enlist the expertise of specialists in the USDA not necessarily available at CDC.
- Required all persons in possession of a Select Agent notify the Secretary of HHS by September 2002.
- Required security risk assessments to be conducted by the FBI of the Department of Justice, in effect security clearances for all individuals who might work with select agents in any laboratory possessing them.
- Required all facilities in possession of select agents to designate an RO who "will need to inventory its facility and consult with others (e.g. principal investigators) as necessary to obtain the information required for this application".[16]

The scope of PL 107-188 was enormous and several iterations of proposed rules for implementing the law were published in the FR. Briefly the official publications (date and FR volume and page) bringing the terms of the law into effect were:

- 7/2/02 (67 FR 44464): Proposed data collection requirements for public comment issued by the CDC.
- 7/12/02 (67 FR 46364): Preliminary Guidance for Notification of Possession of Select Agents issued by the CDC.
- 8/12/02 (67 FR 52383): Interim rule and request for comments, APHIS.
- 12/13/02 (67 FR 76886): Interim final rule ("interim" rules are published as rules that are enforced but which may be changed by subsequent public comment).

---

[15] http://www.ecfr.gov/cgi-bin/text-idx?SID=ad0987e511f9d0bce7cf252d00f17de1&node=42:1.0.1.6.61 &rgn=div5

[16] Federal Register Vol. 67 (127) July 2, 2002: http://www.gpo.gov/fdsys/pkg/FR-2002-07-02/pdf/02-16674.pdf

- 11/3/03 (68 FR 62245): Second interim final rule.
- 3/18/2005: Final rule regarding possession use and transfer of select agents and toxins.

Throughout the 2½-year period – which also included a public meeting in late 2002 – about 150 written comments were received by officials at the CDC and many more at APHIS (the agency within and designated by USDA to implement the requirements of the legislation). Although most of the suggestions submitted by commenters were rejected by the agencies, some clarifications and changes were made to the proposed rules as a result. An additional outcome was a re-organization of 42 CFR 73 to make its structure similar to analogous rules promulgated by the USDA that appeared in 9 CFR 121.[17]

Thus, there are three legislative milestones that have largely defined biosecurity as practiced in laboratories in the United States: The Anti-terrorism and Effective Death Penalty Act of 1996 establishing the first SATL and reporting and archiving requirements for transfer of those agents (later put into effect by the CDC acting on behalf of the Secretary of HHS); the US PATRIOT Act of 2001, defining possession of a biological weapon as a crime, and also "restricted persons" who may not possess or transfer *any* biological agent – that is, not merely "Select Agents" that can cause disease; and PHSBRA that required a biennial review of the SATL by both the Secretary of HHS and the Secretary of Agriculture, and also mandated security requirements for access to listed agents, including background checks (a "Security Risk Assessment") of laboratory researchers and designated RO, performed by the FBI and ultimately approved by the Attorney General.

Recent Executive Orders (EOs) affecting laboratory biosecurity

Presidents may issue EOs in order to "take Care that the Laws [of the United States] be faithfully executed"[18] even though there is no explicit provision for these declarations in the US Constitution. Since the time of George Washington through the administration of Barack Obama more than 13,000 EOs of various types – some proclamations, others directives for implementing laws or establishing policy – have been issued.[19] It is solely at the discretion of the president to determine whether a policy matter or resolution of ambiguities in the administration of laws warrant an EO. So it has been with biosecurity. Shortly before the end of his term in office on January 9, 2009, President George W. Bush issued EO 13486 entitled "Strengthening the Biosecurity of the United States" establishing a formal Working Group of the same name comprised of cabinet secretaries, the Attorney General, the Director of

---

[17] In addition, 7 CFR 331 contains the rules promulgated by USDA for the select agent list for organisms that can damage important food crops. As should now be unsurprising, Title 9 of the Code of Federal Regulations is "Animals and Animal Products", with part 121 addressing "possession, use and transfer of select agents". Title 7 of the Code is "Agriculture" with part 331 also referring to select agents.
[18] US Constitution Article II Section 3.
[19] http://www.presidency.ucsb.edu/data/orders.php

National Intelligence and the Director of the National Science Foundation [19]. Three tasks were assigned to the Working Group:

1. Review and evaluate the efficiency and effectiveness of existing laws, regulations, guidance and practices relating to select agents and toxins, physical, facility, and personnel security and assurance at Federal and non-Federal facilities that function as described above.
2. Obtain information or advice from heads of executive departments and agencies, elements of foreign governments and international organizations with responsibility for biological matters.
3. Prepare a written report to the president, including recommendations for new legislation, regulations, guidance, or practices in laboratories including new oversight mechanisms [4].

In its report in May 2009 the Working Group recommended:

- With respect to Select Agents and Toxins, that the US government perform a risk assessment for each listed item and develop a "stratification scheme that includes biodefense and biosecurity criteria as well as risk to public health."
- With regard to personnel security enhancements to the "Security Risk Assessment ... to allow for improved vetting of US citizens and foreign nationals" with access to select agents and toxins.
- To improve physical security, the development of a "a set of minimum prescriptive security standards."
- A review of existing risk assessment of transportation of select agents and toxins.

Professional societies were quick to respond to the Working Group's recommendations. For example, the American Biological Safety Association (ABSA) commented that: "Select Agent Regulations are [already] sufficiently rigorous" and "should not be made more prescriptive;" that the "Federal Government absolutely should not develop prescriptive physical security requirements;" and that inspections of laboratories with select agents would benefit from "careful selection and training of inspectors," probably reflecting frustration on the part of laboratory managers with questionable results of inspections. There were also concerns that "additional restrictions on shipping will inhibit important research" citing the example of H1N1 samples from Mexico that had to be shipped to Canada because of restricting US import and transfer regulations.[20]

The ABSA also objected to components of enhanced personnel security requirements that would include the "two person rule" in all circumstances and recommended instead that federal funds "be used to develop or enhance existing biosafety and biosecurity training programs." Finally ABSA objected to licensure of individual researchers as "unnecessary and undesirable."

Notwithstanding the criticism of the Working Group's recommendations, EO 13486 was followed about 18 months later by EO 13456 in July 2010. This directive required the Secretaries of HHS and Agriculture to review the SATL in order to:

- designate a subset of the List called "Tier 1 agents" that present "the greatest risk of deliberate misuse with most significant potential for mass casualties" or other severely adverse effects on the economy or public confidence,

---

[20] http://www.absa.org/pdf/090530ABSAcommentsBWG.pdf

- identify options for graded protection of these Tier 1 agents with "tailored risk management practices" and
- consider reducing the number of agents on the SATL.

Thus, since the attacks of September 2001, US presidents have shown increasing interest in biosecurity reflected in two key EOs, the most recent of which directed subject matter experts at CDC and USDA/APHIS to review the SATL, to further stratify the agents based on risk (for greatest potential harm if released), and to make changes in security arrangements at facilities to minimize those risks. We will see shortly how these EOs melded with existing law to generate the most recent modifications to the SATL and other security and procedural requirements in laboratories.

Some important documents from professional and *ad hoc* groups

In addition to the report of the Working Group summarized above, several other reports from biological scientists stand out as key documents providing guidance to the Congress and the executive branch on measures that might be taken to enhance biosecurity in the United States.

The National Research Council in 2003 released "Biotechnology Research in an Age of Terrorism" [20]. This document focused mainly on Dual Use Research of Concern, but also recommended the creation of a National Science Advisory Board for Biodefense (NSABB) that, *inter alia* would periodically review existing federal government legislation to "provide protection of biological materials and supervision of personnel working with these materials." In addition the Council opined "it is crucial to avoid well-meaning but counter-productive regulations on pathogens. Rules for containment and registration of potentially dangerous materials must be based on scientific risk assessment and informed by a realistic appraisal of scientific implications." The Council suggested that the NSABB could "provide advice ... about revising regulations in response to new developments" and that "rules governing transfer of materials between laboratories... might also be regularly reviewed by NSABB" in light of new threats.

Shortly thereafter, the NSABB was chartered by the Secretary of HHS [21].

In 2008, the Commission on the Prevention of Weapons of Mass Destruction, Proliferation, and Terrorism (C-WMD) tasked by the Congress in PL 110-53 (Implementing Recommendations of the 9/11 Commission Act of 2007) produced a comprehensive report entitled "World at Risk" [22]. The Commission regarded as "likely" that there would be a terrorist attack somewhere on the globe utilizing a weapon of mass destruction (WMD) within 5 years, and further believed that terrorists were more likely to use a biological than a nuclear weapon. The Commission's prediction has fortunately not come to pass. Nonetheless, few would regard the risk of WMD use as receding. Among the recommendations of the Commission of importance to biological scientists were that the United States should: "conduct a comprehensive review of the domestic program to secure dangerous pathogens, ... tighten government oversight of high-containment laboratories, ... [and] promote a culture of security awareness in the life sciences community." These measures closely align with the biosecurity principles of access control to disease-causing organisms and toxins, accounting of those agents and personnel reliability programs.

The last of the Commission's recommendations was explored in detail by the NSABB in 2009. In a report entitled "Enhancing Personnel Reliability Among Individuals with Access to Select Agents" [23], board members recognized that research programs involving nuclear materials often involved sensitive national security issues and might serve as a model for addressing the "insider threat" problem in biological laboratories, but at the same time realized that there are significant differences between pathogens and nuclear materials (including nuclear materials that might be part of weapons systems). Noting that local institutions were already successfully screening individuals who might work with select agents (and based on the very few applications rejected by the Department of Justice in the Security Risk Assessment Program), the Board counseled against a "formal, national Personnel Reliability Program" as it might have "unintended and detrimental consequences" for scientific research. Instead the Board suggested that individual institutions should work to enhance a "culture of responsibility and accountability" in combination with education on biosecurity issues within professional societies. The Board also recommended reducing the list of select agents or stratifying them further by risk, which was in fact taken up by EO 13546 in 2010. Thus, we can see the careful considerations of the C-WMD and NSABB finding their way into US government policy.

At about the same time and in response to a request from the Homeland Security Council at the White House, the National Academies of Sciences released its study "Responsible Research with Biological Select Agents and Toxins" that listed a series of recommendations similar to those from the NSABB and also suggested stratifying the list of select agents by risk [24]. The Academy group further advised that some baseline level of physical security be established for facilities holding select agents.

Also in 2009, a US government interagency "Working Group on Strengthening the Biosecurity of the United States" created under EO 13486 reviewed procedures at facilities possessing Select Agents and Toxins, specifically:

- Evaluating the efficiency and effectiveness of existing laws, regulations ... and practices relating to physical, facility and personnel security and assurance at facilities.
- Obtaining advice from heads of US government executive agencies and departments and elements of foreign governments and international organizations that have similar responsibilities.

Like the C-WMD, the Working Group recommended a reduction or stratification of select agents so that security measures might be tailored to the level of risk; require at the local level that facility managers review the behavior and practices of individuals with access to select agents; and development of a set of minimum security standards with enhancements based on risk associated with select agents [4].

All of these reports ultimately contributed to the content of the 2010 EO 13546, "Optimizing the Security of Biological Select Agents and Toxins in the United States," which as noted earlier called for a special designation for the highest-risk agents as "Tier 1 agents," and graded protection of the items on the SATL.

Legislation, deliberation and executive exhortation become regulation

The timing of EO 13456, early July of 2010, coincided with the second Biennial Review of the SATL and an "Advanced NPRM and request for comments" appeared in the FR on 7/21/10 (75 FR 42363) that included the new "tiering" as directed by the president.

As a result of EO 13456 and acting on authorization already enshrined in law as a result of the PHSBPRA of 2002, over the next 2 years the CDC (acting on behalf of the Secretary of HSS) and the APHIS (acting on behalf of the Secretary of Agriculture) produced a final rule[21] on October 5, 2012 that established "Tier 1" Agents as follows[22]:

- Human disease agents and toxins (CDC)
  - Ebola virus,
  - *Francisella tularensis*,
  - Marburg virus,
  - Variola major virus,
  - Variola minor virus,
  - *Yersinia pestis*,
  - Botulinum neurotoxin,
  - Botulinum neurotoxin producing species of *Clostridium.*
- Animal disease agents (APHIS)
  - foot and mouth disease virus,
  - Rinderpest virus.
- "Overlap" agents (CDC and APHIS)
  - *Bacillus anthracis*,
  - *Burkholderia mallei*,
  - *Burkholderia pseudomallei*.

And, some agents and toxins were removed from the SATL. Specifically,

- at the recommendation of the CDC:
  - Herpesvirus 1 (Herpes B),
  - *Clostridium perfringers* toxin,
  - *Coccidioides* species,
  - Eastern equine encephalitis virus,
  - Flexal virus,
  - West African clade of Monkeypox,
  - *Rickettsia rickettsii*,

---

[21] 77 FR 61084

[22] As the SATL undergoes periodic revision, the list produced in the rule-making process of October 2012 is not necessarily the same as the current one that may be found at: http://www.selectagents.gov/SelectAgentsandToxinsList.html [last accessed 25.05.15]

- all conotoxins except the short paralytic alpha conotoxins,
- Venezuelan Equine Encephalitis Virus (subtypes ID and IE) were also removed from the "overlap" select agent category.
- At the recommendation of APHIS:
  - Akabane virus,
  - bluetongue virus (exotic), bovine spongiform encephalopathy agent,
  - camel pox virus,
  - *Ehrlichia ruminantium* (heartwater),
  - Japanese encephalitis virus,
  - malignant catarrhal fever virus (Alcelaphine herpes virus type 1),
  - Menangle virus,
  - Vesicular stomatitis virus (exotic): Indiana subtypes VSV–IN2, VSV–IN3.

Finally, three viruses were added to the SATL with the publication of the new rule: SARS-CoV (the organism causing the severe acute respiratory syndrome first identified in 2003), Lujo virus (a hemorrhagic fever virus closely related to Old World arenaviruses) and Chapare virus (a New World arenavirus also causing hemorrhagic fever in humans).

For Tier 1 agents, new requirements emerged from the CDC and APHIS as follows (with their location in the CFR noted for ease of reference):

- Clinical laboratories must immediately report by telephone, facsimile or e-mail the identification of any of the HHS or overlap Tier 1 agents (42 CFR 73.5 and 73.6).
- Security plans for select agents must also include for Tier 1 agents (42 CFR 73.11):
  - A description of pre-access suitability assessment of persons who will have access to Tier 1 agents.
  - Procedures for how an RO will coordinate their efforts with the entity's safety and security professionals.
  - Procedures for ongoing assessment of the suitability of personnel with access to Tier 1 agents.
- Additional security enhancements must be put in place to (42 CFR 73.11):
  - Limit access to Tier I agents only those individuals who are approved by the HHS Secretary after FBI security risk assessment.
  - Outside of normal laboratory hours, RO approval access to agents.
  - A minimum of three security barriers where each barrier adds to the delay in reaching secured areas where agents are stored.
- Facility biosafety plans must include an occupational health program for individuals with access to Tier 1 agents (42 CFR 73.12).
- Facility incidence response plans must describe an entity's response in the case of failure to detect intrusion and procedures for notifying appropriate officials and law enforcement (42 CFR 73.14).

Hence, four related but distinct forces have resulted in current laboratory biosecurity practices and procedures: statutes, landmark studies from practicing professionals, EOs from the president, and recommendations from practitioners via professional organizations, academia and individual comment (see Figure 1.1). It is important to note that via the rule-making process subject matter experts in laboratory practices and management had the opportunity to influence the formulation of the SATL and

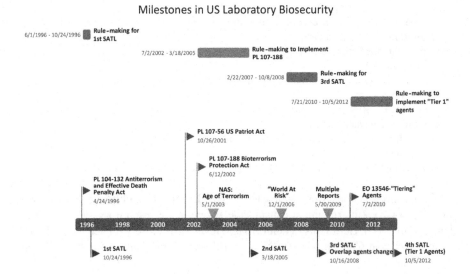

Figure 1.1 Biosecurity timeline. SATL, Select Agent/Toxin List; PL, Public Law; EO, Executive Order; NAS, National Academy of Sciences.

some of the physical security and accounting practices now in force at facilities holding those potentially dangerous materials.[23]

The formation of the SATL and its associated security measures is a good template for gaining insight into federal agency rule making after a new law is enacted. Since the comments received by an agency can be quite extensive, they may overwhelm the ability of individuals reviewing them to do so within the 30- or 60-day comment period. In addition, some comments may point to new information that was either overlooked or not available at the time of the original FR publication. Thus, the rule-making period may be extended; notification of such is made via the FR.

The evolution of the SATL – the centerpiece of much of the biosecurity apparatus in the hundreds of facilities across the United States that hold the agents – is summarized in the timeline from 1996 through 2012 when the "tiering" of select agents came into force as a result of EO 13546 (see Figure 1.1). It appears clear that biosecurity rule-making has, in the recent timeframe, become very complex and may extend over several years, involving as it does a wide variety of Congressional mandates and the need to solicit comments from researchers in academia and industry working with dangerous pathogens. Whether or not the plethora of regulations will result in actually strengthening biosecurity remains to be seen.

---

[23] Professionals in the field are not always successful in changing the initial draft of an agency's proposed rules. In this particular episode of rule-making, 30 comments were received "from researchers, scientific organizations, laboratories and universities" (see 77FR 61056, dated October 5, 2012, accessible at: http://www.gpo.gov/fdsys/pkg/FR-2012-10-05/pdf/2012-24434.pdf). Most of the recommendations for changes in the proposed rule were rejected by, in this case, the Department of Agriculture, but the rationale for each rejection was specifically addressed. In other cases, recommendations from comments are more frequently incorporated into the Final Rule.

What might the future bring?

As of the time of this writing, there have been two recent events that are likely to result in yet more oversight of biosecurity practices in the United States.

On June 6, 2014, CDC researchers working in the Bioterrorism Rapid Response and Advanced Technology (BRRAT) laboratory transferred samples of *B. anthracis* (Ames strain) from a Biosafety Level 3 (BSL-3) suite to a BSL-2 lab on the belief that neither viable vegetative cells nor spores were in the samples as they had been subjected to an inactivation process before being moved. However, at least some of the samples were *not* sterile, potentially risking infection in CDC personnel, although subsequent investigation revealed this possibility to be "highly unlikely" [25]. This breech of biosafety practice was announced on July 19, 2014 [26].

Further, on July 8, 2014 the CDC reported that several vials labeled "variola" (variola virus is the causative agent of smallpox) had been found at a cold storage facility operated by the Food and Drug Administration on the campus of the National Institutes of Health (NIH) in Bethesda, Maryland [27]. Subsequent testing performed at the CDC confirmed the presence of viable virus in two of six vials.[24] By international agreement, all variola virus is to be kept in secure storage at two laboratories: at the CDC in Atlanta and at the State Research Center for Virologic Research in Koltsovo, Russia, and is technically "owned" by the WHO, which must approve all experiments involving its use. While such news was not wholly surprising – variola had turned up in at least two European laboratories since the WHO required destruction of the virus or transfer to the official WHO repositories – it did raise the question of accountability for extraordinarily dangerous pathogens in the United States.

Also, in January 2014 at the CDC's Influenza Division laboratory in Atlanta, a sample of low-pathogenic avian influenza type A (subtype H9N2) was contaminated with a highly pathogenic avian influenza type A (subtype H5N1) with "subsequent shipment of the contaminated culture to an external high-containment laboratory." Although there were no apparent adverse effects, the episode clearly posed a risk to individuals in the receiving laboratory (and perhaps others who handled the material before it was shipped). Many new review procedures and oversight processes were put into place (presumably at substantial cost) at the CDC [28].

These episodes were more than an embarrassment for the NIH and CDC. Since the CDC provides the staff to carry out inspections at facilities registered to hold Select Agents and Toxins, most researchers would probably assert that the CDC has a special responsibility to uphold all biosafety and biosecurity regulations. The head of the BRRAT resigned. Perhaps even worse, the United States was in violation of a critical international agreement intended to prevent the reintroduction of variola into the human population.

At the end of July, CDC Director Frieden announced the formation of an external advisory group comprised of leading researchers and biosecurity experts to "provide

---

[24] The vials labeled "variola" were co-located with a total of 327 vials in 12 boxes some of which were labeled as containing other possible SATL agents such as *Ricketssia* sp. See http://www.fda.gov/NewsEvents/Newsroom/PressAnnouncements/ucm405434.htm

advice and guidance to the CDC Director and ... Director of Laboratory Safety" [29]. It remains to be seen whether the advisory group's recommendations will lead to further revision of 42 CFR 73 and 9 CFR 121 biosecurity regulations.

Many researchers have noted that well-intentioned regulations designed to protect public safety and limit the illicit use of select agents result in unintended consequences including significant costs to bring facilities into compliance with technical requirements [30] and decreased productivity as measured by publication count and even abandonment of research involving live agents [31].

Summary

Current-day biosecurity procedures and regulations are, in large measure, a result of several near-calamitous terrorist events in the United States, raising public awareness of the threat posed by a particular infectious agents and toxins that, if released by intent or accident, could have devastating effects on human or animal health, and even that of important food plants. Congress has responded by passing legislation that imposes formidable requirements on laboratories possessing "select agents and toxins," and the executive branch of government has, as compelled by the Constitution, implemented those mandates with the assistance of technical experts inside and outside of government. Substantial but hard-to-quantify costs accrue to facilities that in turn may decrease research productivity.

It is probably impossible to know whether existing biosecurity legislation passed starting in 1996 actually reduces the chances of illicit use of biological materials or even accidents in their transfer or handling. Recent security breeches will likely result in additional Congressional action and EOs, some of which may prove costly. Laboratorians have opportunities to influence the process of rule-making by which laws are brought into effect, but only with a keen awareness of the process and review of proposed regulations in order to provide key federal agency administrators with critiques and suggestions to reduce the prospect of new biosecurity requirements that are either ineffective or onerous. While there is no guarantee that intervention from professionals will stem the tide of regulations, laboratory managers, biosafety officers and institutional oversight committee members must make an effort to remain current with proposed rules and to vigorously comment to regulatory agencies.

References

[1] Ragland D. Overview of biosecurity. In: Aiello SE, editor. 2012.
[2] Convention of the prohibition of the development, production and stockpiling of bacteriological (biological) and toxin weapons and on their destruction. In: United Nations, editor. 1975.
[3] Spokes PJ, Ferson MJ, Ressler KA. Staff immunisation: policy and practice in child care. J Paediatr Child Health 2011;47:530–4.

[4] Office of the Assistant Secretary for Preparedness and Response. In: Report of the Working Group on Strengthening the Biosecurity of the United Stated. US Department of Health and Human Services, editor. Washington, DC; 2009.

[5] Centers for Disease Control. Biosafety in microbiological and biomedical laboratories. In: US Department of Health and Human Services, editor. 2009. Health and Human Services Department, HHS Publication No. 21-1112.

[6] Casadevall A, Relman DA. Microbial threat lists: obstacles in the quest for biosecurity? Nat Rev Microbiol 2010;8:149–54.

[7] Hottes AK, Rusek B, Sharples FR. Biosecurity challenges of the global expansion of high-containment biological laboratories. Istanbul: National Academies Press; 2011.

[8] Berg P, Baltimore D, Brenner S, Roblin RO, Singer MF. Summary statement of the asilomar conference on recombinant DNA molecules. Proc Natl Acad Sci USA 1975;72:1981–4.

[9] Howard-Jones N. The scientific background of the Internatioal sanitary conferences, 1851–1938. Geneva: World Health Organization; 1975.

[10] League of Nations Bulletin of the hygiene organization of the league of nations, 1920–1945 (in French). Geneva: League of Nations; 1945.

[11] Ryan ET. Eyes on the prize: lessons from the cholera wars for modern scientists, physicians, and public health officials. Am J Trop Med Hyg 2013;89:610–4.

[12] World Health Organization. Laboratory biosafety manual. 3rd ed. In: World Health Organization, editor. Geneva: 2004.

[13] Carr K, Henchal EA, Wilhelmsen C, Carr B. Implementation of biosurety systems in a Department of Defense Medical Research Laboratory. Biosecur Bioterror 2004;2:7–16.

[14] Beall JA. Are we only burning witches? The antiterrorism and effective death penalty act of 1996's answer to terrorism. Indiana Law J 1998;73 (Article 17).

[15] Danzig R, Sageman M, Leighton T, Hough L, Yuke H, Kotani R, Hosford ZM. Aum Shinrikyo: Insights into how terrorists develop biological and chemical weapons. Washington, DC, USA: Center for a New American Security; 2014.

[16] The Antiterrorism and Effective Death Penalty Act. 1996.

[17] Uniting and Strengthening America by Providing Appropriate Tools Required to Intercept and Obstruct Terrorism (USA PATRIOT Act) Act of 2001. (Public Law 107-56, October 26, 2001).

[18] U.S. Congress. Public Health Security and Bioterrorism Preparedness and Response Act of 2002. P.L. 107-188. 42 U.S.C. 243, June 12. Available from: <http://tis.eh.doe.gov/biosafety/library/PL107-188.pdf>.

[19] Executive order 13486: Strengthening laboratory biosecurity in the United States. In: Federal Register; 2009:2289-2291.

[20] Biotechnology research in an age of terrorism. Committee of Research Standards and Practices to Prevent the Destructive Application of Biotechnology, Washington, DC: National Academy of Sciences; 2003.

[21] Shea DA. Oversight of dual-use biological research: the National Science Advisory Board for Biosecurity. Washington, DC: Congressional Research Service; 2007.

[22] Graham B, Talent J. World at risk. The report of the commission on the prevention of weapons of mass destruction proliferation and terrorism. New York, NY: Vintage Books; 2008.

[23] Enhancing personnel reliability among individuals with access to select agents. National Science Advisory Board for Biosecurity (NSABB); 2009.

[24] National Research Council (NRC). Responsible research with biological select agents and toxins. Washington, DC: National Research Council; 2009.

[25] Centers for Disease Control. Report on the potential exposure to Anthrax 2014, 7/25/14.
[26] Media Statement. CDC Lab determines possible anthrax exposures: staff provided anti-biotics/monitoring. Available from: <http://www.cdc.gov/media/releases/2014/s0619-anthrax.html>.
[27] Reardon S. Infectious diseases: smallpox watch. Nature 2014;509:22–4.
[28] Centers for Disease Control. Report on the inadvertent cross-contamination and shipment of a laboratory specimen with influenza virus H5N12014.
[29] Media Statement. CDC announces the formation of an external laboratory safety work-group. Available from: <http://www.cdc.gov/media/releases/2014/s0724-lab-workgroup.html>; 2014 [accessed 10.08.14].
[30] Matthews S. Select-agent status could slow development of anti-SARS therapies. Nat Med 2012;18:1722.
[31] Dias MB, Reyes-Gonzalez L, Veloso FM, Casman EA. Effects of the USA PATRIOT Act and the 2002 bioterrorism preparedness act on select agent research in the United States. Proc Natl Acad Sci USA 2010;107:9556–61.

US federal oversight of biological materials and the IBC

Deborah Howard

Chapter Outline

Introduction

Research in life sciences in the last century has provided advances in agriculture and industrial development while transforming the practice of medicine. In the field of biopharmaceutical products, major breakthroughs have been made including human recombinant insulin for the treatment of diabetes, a vaccine against hepatitis B, and medicines for cancer therapy, arthritis, multiple sclerosis, and cystic fibrosis [1].

Life science refers to the study of living organisms including, microbes, human beings, animals, fungi, and plants. It includes the fields of biology, aerobiology,

Ensuring National Biosecurity. DOI: http://dx.doi.org/10.1016/B978-0-12-801885-9.00002-0

agricultural science, plant science, animal science, bioinformatics, genomics, prot-
eomics, synthetic biology, environmental science, public health, modeling, engineering of living systems, and many other types of scientific study [1].

Better understanding of the principles of genetics, biochemistry, and the structure
of DNA, as well as the discovery of gene-splicing technology have greatly advanced
biological or life sciences. The discovery of genetic engineering was a major accomplishment for the scientific community. Genetic engineering allows for the modification and transfer of genetic material from one organism to another including from one
species to another. These scientific advances are directly attributable to the sharing of
knowledge among scientists and people with ideas moving freely between universities,
government agencies, and private industry [1].

The early days of recombinant DNA research

In the late 1960s, biochemists and molecular biologists made remarkable progress in
the study of DNA, RNA, and enzymes that are part of the replication process. These
scientific advances caused the public and scientific community anxiety regarding
the potential hazards of these cutting edge discoveries. If viruses could be made in
the laboratory, there was concern for what could happen if they were pathogenic to
humans or animals. Public fears about this new technology at the time were fed by
Michael Crichton's book *The Andromeda Strain* [2].

Recombinant DNA (rDNA) technology is the insertion of genetic material from
one organism into the genome of another organism. This genetic material is then rep-
licated and expressed by the receiving organism. The invention of rDNA technology is
largely attributed to Drs Paul Berg, Herbert Boyer, and Stanley Cohen; however, many
other scientists have also made important contributions to this field of science [3].

Paul Berg was a biochemist at Stanford University who was interested in the genetics of microbes. He wondered if it was possible to insert foreign genes into a virus,
causing it to become a vector that would carry those genes to new cells. In 1971, Berg
conducted a groundbreaking, controversial, gene-splicing experiment that involved
splicing Simian Virus 40 (SV40) with an *Escherichia coli* restriction enzyme. Berg's
gene-splicing experiment resulted in the insertion of the restriction enzyme into
virus-infected cells. This was the first man-made rDNA; as such experiments were
eventually referred to in the research field [3]. This experiment eventually led to the
inception of the federal regulation of rDNA.

In 1972, Herbert Boyer from the University of California at San Francisco, in collaboration with Stanley Cohen from Stanford University, advanced the development
of modern biotechnology by inserting rDNA into bacteria where it would replicate
naturally [3]. These experiments led to the breakthrough synthetic development of
somatostatin, the hormone that plays a major role in regulating growth hormone and
the hormone insulin that enables cells to absorb glucose from the blood [3].

Berg did not immediately move forward in his research inserting the rDNA into
another organism because of the public concern over the possible risk of such experiments [3]. Berg's experiment was considered controversial because SV40 is a monkey
virus that can transform human as well as nonhuman primate cells into a cancerous

state. Because of the perceived risks, Berg postponed his research and, to begin the process of evaluating the potential hazards, he organized a conference to discuss biohazards in biological research in January 1973 at the Asilomar Conference Center in Pacific Grove, California. This was the first conference that focused on laboratory safety and design to protect laboratorians handling rDNA. Following this meeting, the National Institutes of Health (NIH) and National Institute of Medicine (NIM) were asked to appoint a committee to study this new technology. Subsequently, a Recombinant DNA Advisory Committee (RAC) was set up by the NIH at the request of the scientific community to review recombinant experiments [2].

The RAC is a federal advisory committee that provides recommendations to the NIH director as they relate to basic and clinical research involving recombinant or synthetic nucleic molecules. All RAC proceedings and reports since 1999 are posted on the NIH's Office of Biotechnology Activities (OBA) website, publicly available to the scientific and lay communities alike, and thereby promoting transparency and accessibility to basic and clinical research involving recombinant or synthetic nucleic acid molecules [4].

OBA is the governing body that oversees rDNA research and the RAC. OBA promotes science, safety, and ethics in the advancement of public policies in three different areas: Biomedical Technology Assessment, Biosafety, and Biosecurity. The office is tasked with developing policies related to (i) conduct of clinical trials when using recombinant and synthetic nucleic acids at an institution receiving NIH funding, (ii) biosafety for those receiving NIH funds, (iii) biosecurity, including oversight of dual-use research of concern (DURC) for those institutions receiving NIH funding, and (iv) registration of new stem cell lines for institutions receiving NIH funding. OBA also updates and interprets biosafety policies under the NIH guidelines involving research with recombinant and synthetic nucleic acids. Their input is critical for emerging governmental policies, such as the recently published regulations for DURC [5].

As a result of rDNA research advances and safety concerns, in 1975 Berg organized another and more in-depth conference at Asilomar focused on establishing voluntary guidelines to ensure the safety of working with rDNA. It was a three-and-a-half-day meeting attended by both scientists and members of the press. Approximately 150 of the leaders in this emerging field debated the risks of cloning, manipulating foreign genes and their expression in bacteria. At the end of this meeting, a series of resolutions that set forth guidelines for physical and biological containment procedures was given to the newly established NIH RAC. RAC became the governing body for the implementation of the guidelines. The public participated in the design of the guidelines through open hearings, and the guidelines were given strength by linking compliance with federal funds for any such research. The NIH guidelines established roles and responsibilities for institutions including implementing policies for the safe conduct of research subject to the NIH Guidelines and establishing an Institutional Biosafety Committee (IBC) to oversee rDNA research at the local level in individual laboratories [6].

It is interesting to note the experiments that were prohibited under the Asilomar Guidelines in 1975 are still regulated in the latest 2013 edition of the NIH Guidelines [7], and therefore, by IBCs at the local level. The experiments prohibited at the Asilomar conference were the following: (i) transfer of drug resistance traits not found in nature that would affect control of disease, (ii) deliberate formulation of rDNAs containing genes for the biosynthesis of toxins of very high toxicity, (iii) deliberate creation of rDNAs that can increase virulence or host range, (iv) release of rDNAs into the environment, and (v) large-scale experiments – more than 10 liters of culture. Even in 1975, the participants at the Asilomar conference recognized these experiments could be used for malevolent or dual-use purposes in life science research [6].

In the most recent edition of the NIH Guidelines, the most regulated rDNA experiment is still the transfer of drug-resistant traits that would affect the control of disease. In order to conduct this experiment, the IBC must request the OBA to make a determination regarding whether an experiment involving the deliberate transfer of a drug-resistant trait falls under NIH Guidelines Section III-A-1-a and therefore requires RAC review and NIH Director approval before experiments can begin [7].

IBCs were federally mandated in the NIH Guidelines for Research Involving Recombinant DNA Molecules for institutions that receive federal funds to conduct rDNA research. At the local level, IBCs are the foundation of oversight for research involving recombinant and synthetic nucleic acids. The IBC's primary responsibility is to safeguard protection of the personnel, public, and environment as it pertains to rDNA technology experiments [8].

In March 2013, the NIH Guidelines were updated to include synthetic nucleic acid molecules. This change was implemented for two reasons: (i) Recognition of the correct biosafety containment level of an agent is important regardless of the technology used to generate the agent (i.e., recombinant or synthetic methods). (ii) The National Science Advisory Board for Biosecurity (NSABB) recommended that the US Government partner with the scientific community to ensure that current biosafety guidelines are appropriate, adequate, and easily understood with respect to working with synthetic nucleic acids [7].

The new language of the Section I-A of the NIH Guidelines states "The purpose of the NIH guidelines is to specify the practices for constructing and handling: i) recombinant nucleic acid molecules, ii) synthetic nucleic acid molecules, including those that are chemically or otherwise modified but can base pair with naturally occurring nucleic acid molecules, and iii) cells, organisms and viruses containing such molecules." Within the NIH Guidelines, the term "recombinant DNA molecules" has been changed to "recombinant or synthetic nucleic acid molecules" to include research with recombinant and synthetic nucleic acids [7].

Components of an IBC

According to the NIH Guidelines an IBC has to consist of no less than five members with the applicable recombinant or synthetic nucleic acid molecule technology and be able to evaluate the safety of recombinant or synthetic nucleic acid molecule research and recognize any potential risk to public health or the environment. At least two members must be community members not associated with the institution. If the institution conducts plant or animal research, experts specializing in this field must be represented on the committee. If the institution produces large-scale recombinant research material (>10L) or works at high containment levels (such as biosafety level 3 or 4), a Biological Safety Officer (BSO) must be appointed and serve on the IBC. The role of the BSO is to develop emergency plans for handling accidental spills and personnel contamination with recombinant or synthetic DNA, and they must report any recombinant or synthetic nucleic acid problems, violations, and research-related injuries or illnesses to the IBC. In addition, the BSO advises on security and offers technical advice to the Principal Investigators and IBC on safety procedures. The committee must have expertise in biological safety, containment, risk assessment, applicable law, and knowledge of institutional policies. As previously mentioned, the NIH Guidelines require at least two community members to be on the committee. Community members on the IBC are an important component, because they bring an outside perspective with respect to health and the protection of the environment. Community members are usually representatives from local public health or environmental authorities, and are frequently members with laboratory, medical, occupational, or environmental experience. The community members represent the concerns of the local population. When experiments contain topics that fall outside the knowledge of the IBC committee, it is appropriate to seek *ad hoc* members with appropriate additional expertise. IBCs are required to register with the OBA and file an annual online update. The update must identify the Chairman and contain contact information and biographical sketches for all members. The BSO can be the chairman of the committee if they have the relevant scientific background to evaluate the research protocols [8].

While the IBC was originally formed to review research with rDNA for conformity with the NIH guidelines, the committee was also tasked with conducting risk assessments to protect the environment and public health. When the NIH guidelines were established, the IBC evaluated containment levels using the NIH Guidelines as a guidance tool [2]. Assessing the adequacy of the facilities, reviewing standard operating procedures (SOP) and training were also under the original responsibilities of the IBC [5]. Other roles assigned to the IBC are the enforcement of institutional and investigator compliance along with serious adverse event (SAE) reporting to regulatory bodies such as the RAC and local oversight such as the Institutional Review Board and the Institutional Animal Care and Use Committee (IACUC). As the NIH regulations have evolved, IBCs were also given the role of reviewing human gene therapy protocols to certify compliance with Appendix M of the NIH Guidelines verifying that participants were not enrolled in a clinical trial until the appropriate approvals had been obtained [7]. In many cases, smaller institutions without the committee expertise to review human gene therapy protocols have to resort to hiring an outside IBC for this purpose.

All human gene transfer clinical trials occurring at or sponsored by institutions receiving NIH funds for research with recombinant or synthetic nucleic acid molecules have to be submitted to OBA for review by the RAC. RAC review must be completed before local IBC or IRB review. SAEs from human gene protocols were also given to the IBCs to oversee, as previously stated [8]. An SAE is any detrimental outcome associated with the use of a medical or biological product administered to a patient. In the case of an adverse event being reported and reviewed by the IBC, the incident has to involve recombinant or synthetic nucleic acid molecules. The event is "serious" and is required to be reported to FDA when the patient outcome is death, disability or permanent damage, life threatening, or hospitalization as a result. Emergency room visits that do not result in hospital admission should also be documented and evaluated by the IBC to determine whether it is an adverse event [9].

The role of the IBC has evolved in the last 30 years to review much more than rDNA projects. Many IBCs are now charged with reviewing the use of select agents and toxins, bloodborne pathogens, xenotransplantation, DURC, nanotechnology, and synthetic nucleic acid research. Some IBC oversight also includes reviewing policies and procedures and facility review for containment. With the new personnel suitability requirements of the Federal Select Agent Program (SAP), some institutions working with select agents are now requiring their IBCs to review biosecurity and personnel suitability plans before institutions adopt them. The increase in regulatory burden has significantly increased the workload for this committee of volunteers. Research projects involving recombinant and synthetic nucleic acid molecules at institutions have increased significantly in the last 15 or so years. The budget for the NIH has more than doubled from $13.7 billion in 1998 to $31.3 billion in 2014. Due to world events, IBCs have had to expand their oversight responsibilities to include biodefense and emerging infectious disease research review. New technological capabilities are also of concern to IBCs, such as the genomic synthesis of viruses (e.g., polio), reconstructing organisms (e.g., the 1918 pandemic influenza strain), and novel gene therapy protocols [8]. The ease of rDNA techniques and access to necessary materials (along with the ability to purchase genetics elements off the internet) has increased the need for tighter oversight and restrictions on the publication of life science research.

Since the 1990s, the debate about the possible control or censorship of research in life science research has intensified. The timeline for concern began with the terrorist attacks on September 11, 2001 and the anthrax letters in the United States mail the following week. These events were followed by the 2001 report in the *Journal of Virology* regarding a re-engineered mousepox virus intended to sterilize mouse populations that unexpectedly produced a much more virulent virus [10]. In *Science* in 2002, researchers reported they reconstructed poliovirus from chemically synthesized oligonucleotides linked together and transfected into cells [11]. That same year in *Proceedings from the National Academy of Sciences*, researchers reported on the modification of the immune response under the influence of a virulence gene from vaccinia (vaccine strain for smallpox), including information on how to increase viral virulence [12]. All of these DURC experiments raised fears about their potential for bioterrorism.

Due to the threat of bioterrorism, over the years Congress has passed legislation requiring US Government agencies to promulgate rules and regulations to regulate life science research to protect public health and enhance national security. When new laws and regulations are disseminated, an established body that understands the science must have oversight to assure proper implementation. Over the years, this burden of oversight has, by default, fallen to the IBC at the institutional level.

Regulatory agencies and oversight

The regulatory structure for life sciences research has evolved over the last five decades. Responsibility for regulation in the fields of biotechnology research in life sciences in the United States now falls under the jurisdiction of several federal agencies. Over the years various biological incidents including mishandling, releases, illnesses, and accidents have occurred throughout the world. The most recent incidents in 2014 at the CDC include the mishandling of the bacteria anthrax, the Ebola, and the H5N1 viruses. These events have forced the CDC to change the agency's overall safety culture as well as develop and implement stronger oversight measures in order to restore the public's confidence in the CDC [13]. Over the years, as a direct result of public pressure, federal agencies have responded to these incidents by publishing additional guidelines and regulations to control the use and to monitor access to the biotechnology. IBCs have been required to adopt and adhere to these regulations and guidelines in order to conduct research at their institutions.

National institutes of health, environmental protection agency, and federal drug administration

The NIH and other national and international agencies support the publication of federally funded, fundamental research. Fundamental research is defined as basic and applied research in the areas of science and engineering where the resulting information is intended to be published and shared broadly within the scientific community with no governmental restrictions [14]. The technology used during the research is publicly available and may even be a part of the published information [14].

The NIH is the original governing regulatory agency for rDNA and the driving force behind implementing the IBC committee at the local level to oversee rDNA research at an institution. Standards and procedures are set for all NIH-funded research involving rDNA, which the IBC must adopt and follow. Research involving human gene therapy is a special subcategory under the NIH guidelines. Both the NIH and FDA are required to review the research protocols before the initiation of the research. If recombinant material will be released to the environment for crop improvement or other environmental applications, involvement of the Environment Protection Agency (EPA) is also required. Human research and environmental release have added to the administrative and reporting requirement burden on the IBC as they must also review the protocols at a local level to ensure compliance with biosafety and containment requirements [8].

IBC oversight of the SAP

Biological Select Agents and Toxins (BSAT) are agents and toxins determined by the Federal Government to have the "potential to pose a severe threat to public health and safety and animal or plant health and products" [15]. To advance scientific knowledge regarding biological agents and toxins while increasing knowledge of biological countermeasures, academic, commercial, and government institutions have been authorized by the US Government to perform research using these agents [16]. The oversight of the SAP has fallen to the IBC at many organizations. Some IBCs are now required to review SOPs accident and illness reports along with incidents involving theft, loss or release of BSAT agents and ensure the facilities will contain the agent with no release outside of the laboratory. Although not required by the select agent regulations, the annual review of biosecurity plans involving the inventory and security of select agents has also fallen under the purview of the IBC at some institutions.

No government program for the oversight of BSAT existed in the United States before 1996. In that year, Congress passed the Antiterrorism and Effective Death Penalty Act following an incident involving a person without a research need that ordered plague strains from a supplier of biological agents. After the arrest, government officials realized they had no legal right to charge the individual with a crime other than mail fraud [17]. The Act authorized the Secretary of the Department of Human and Health Services (DHHS) to regulate the transfer of BSATs harmful to humans. DHHS requested the CDC to develop regulations that would manage BSAT to protect the public without hindering scientific research. As a result, the CDC was designated as the agency within DHHS responsible for enforcing this regulation [17].

Select agent oversight is a shared federal responsibility between the DHHS, US Department of Agriculture (USDA) and Department of Justice (DOJ). Congress authorizes DHHS to regulate the possession, use and transfer of BSAT. The Secretary of DHHS assigned this authority to the CDC. Congress provided USDA/APHIS the authority to regulate BSAT that pose a severe threat to animal and plant health and/ or products respectively, while the DOJ is responsible for conducting background checks (aka Security Risk Assessments) of individuals that have access to laboratories performing BSAT research [16].

Uniting and Strengthening America by Providing Appropriate Tools Required to Intercept and Obstruct Terrorism (USA PATRIOT Act) Act of 2001

In 2001, Congress bolstered the SAP by passing the *Uniting and Strengthening America by Providing Appropriate Tools Required to Intercept and Obstruct Terrorism (USA PATRIOT ACT) Act of 2001* [18]. This expanded the regulations by restricting the shipping, possession, and receipt of BSAT. The USA PATRIOT Act established requirements for the appropriate use of BSAT. It also identified individuals who should be restricted from working with these agents and imposes criminal and civil penalties for the inappropriate use of BSAT [17]. As a direct result of implementing the PATRIOT Act, as it pertains to BSAT rDNA research, it provided more authority to the IBC and the BSO on the oversight of the BSAT program.

U.S. Congress. Public Health Security and Bioterrorism Preparedness and Response Act of 2002

In 2002, Congress passed another act called the *Public Health Security and Bioterrorism Preparedness and Response Act* [19] which significantly improved oversight of BSAT. This act strengthened the regulatory authority of DHHS under the Antiterrorism and Effective Death Penalty Act of 1996 by requiring a security risk assessment be conducted for individuals having access to BSAT and granted similar regulatory authority to USDA/APHIS for BSAT that present a significant threat to animal or plant health and/or products. It also required coordination and agreement between DHHS and USDA/APHIS on program activities such as the development of regulations, reporting forms, and approval of changes to regulated laboratories' registrations and inspection process for BSAT regulated by both agencies [17].

Over the years, the Bioterrorism Act has been augmented through a series of additional regulations. The DHHS published an interim final rule, the "Possession, Use, and Transfer of Select Agents and Toxins" Interim Final Rule (42 CFR 73, 9 CFR 121, and 7 CFR 331) (effective on February 7, 2003), which implemented the relevant provisions of the Bioterrorism Act. These rules became effective on April 18, 2005. On October 20, 2005, DHHS established an Interim Final Rule adding reconstructed replication competent forms of the 1918 pandemic influenza virus containing any portion of the coding regions of all eight gene segments to the DHHS select agent list. These regulations are referred to as the "Select Agent Regulations." The regulations were recently updated and implemented on October 5, 2012 [20].

National Research Council (NRC)

The NRC is an operating arm of the United States National Academies with the mission of improving government decision making and public policy. Their goal is to increase public understanding while educating and disseminating information in the fields of science, engineering, technology, and health. The council publishes independent expert reports to inform the US policy-making process. Their objective is to improve the lives of people around the world [21]. The NRC's "Committee on Research Standards and Practices to Prevent the Destructive Application of Biotechnology" published the book, *Biotechnology Research – An Age of Terrorism: Confronting the Dual Use Dilemma*" in 2004. This was the first report that specifically addressed national security and life sciences. In the biosafety community, this book is referred to as the "Fink Report" named after the Chairman of the report Gerald R. Fink [22]. The Fink Report contains seven recommendations to ensure responsible oversight for biotechnology research with potential bioterrorism applications. One of the recommendations was to create the NSABB within the Department of Health and Human Services to educate scientists and advise the government on the oversight of "dual use of concern research" [1]. When first published, the "Fink Report" was distributed by the institutions receiving NIH funding to IBC Chairs, administrators and BSOs to review. The book and its contents were discussed at IBCs around the nation. The book indicated the landscape of oversight for life science research was changing, and the IBC would likely take an active role in the oversight of "DURC" as it pertains to rDNA and life science research.

National Science Advisory Board for Biosecurity (NSABB)

Following the publication of the "Fink Report," the Secretary of the DHHS created the NSABB in 2004. The NSABB responsibilities carried out many of the recommendations suggested by the National Academies. The NSABB is an oversight committee tasked with proposing a framework for the identification, review, conduct, and communication of life sciences research with dual use potential. Their role is to consider both the protection of national security while promoting life science research. The group advocates the free and open exchange of information in the life sciences area. Their aim to address dual use concerns is to raise awareness of the issues while strengthening the scientific culture within the research community by increasing understanding and fostering responsibility. The NSABB recommends strategies and has developed tools for researchers to assist in communicating results of research with dual use concerns [23].

NSABB – US policy for oversight of life sciences DURC 2012 and how it affects IBCs

In the language of the Department of Commerce (DOC) or International Traffic and Arms (ITAR) regulations, dual use refers to technology intended for civil end use while also having a military (or terrorism) application. Technology encompasses more than just a product, it also includes how to produce and use the product [14]. In the life sciences field, the definition of dual use is "biological research with a legitimate scientific goal that could be misused by rogue states, terrorist organizations, or individuals to pose a biological threat to public health or national security" [22].

Dual use potential is inherent in life sciences research. However, research with a smaller subset of BSAT organisms – currently 15 in number – is considered to have a higher likelihood for providing knowledge, technology, or end products that can be used to threaten public health or national security including the greatest risk of deliberate misuse. These agents combined with the seven categories of experiments of concern are currently considered "DURC" [24].

Seven experiments of DURC

The NSABB has identified seven categories of research that fall under the category of DURC. These research projects warrant close scrutiny to the potential nefarious application of the research by the investigator or others when being designed, conducted, and published [24].

These include research with the ability to initiate the following:

1. Enhance the harmful consequences of the agent/toxin (e.g., experiments designed to make seasonal flu as virulent as the pandemic 1918 influenza virus).
2. Disrupt immunity to or the effectiveness of an immunization against the agent/toxin without clinical or agricultural justification (e.g., inserting an immunosuppressive cytokine into a viral genome to make the immune response less effective).
3. Confer agent/toxin resistance to clinically or agriculturally useful prophylactic or therapeutic interventions against that agent/toxin or facilitates their ability to evade detection methodologies (e.g., altering the sensitivity of the bacterium *Yersinia pestis* (plague) to doxycycline, the antibiotic used to treat plague infections).

4. Increase the stability, transmissibility, or ability to disseminate the agent or toxin (e.g., genetically modifying a pathogenic thermophilic bacterial strain to grow at ambient temperatures).
5. Alter the host range or tropism of the agent/toxin (e.g., altering a mosquito-transmitted virus so that it is transmissible via a new species of mosquito).
6. Enhance the susceptibility of a host population to the agent/toxin (e.g., modification of a pathogen to allow it to evade a crucial host immune response).
7. Generate or reconstituting an eradicated or extinct agent/toxin (e.g., 2005 reconstruction of the 1918 pandemic influenza virus) [24].

2012 – Current events on dual use research

In late 2011, two federally funded research groups led by Drs Ron Fouchier (Erasmus Medical Centre, Rotterdam, Netherlands) and Yoshihiro Kawaoka (University of Tokyo, Japan and University of Wisconsin-Madison, United States) submitted manuscripts detailing methods for increasing human-to-human transmissibility of H5N1 avian influenza virus (bird flu). These experiments are referred to as "gain-of-function" research because the experiments alter the pathogens in ways that give them features or functions not found in nature or in the wild. Both groups genetically modified highly pathogenic, avian influenza H5N1 viruses, some through re-assortment with the H1N1 (swine) influenza virus (Kawaoka experiment), directed mutation (Fouchier), and virus passage experiments (Fouchier). The resulting viral mutants could be transmitted via aerosol among ferrets, a model for influenza transmission in human beings [25]. The submission of these two manuscripts to the scientific journals *Nature* and *Science* caused pandemonium in the scientific community and resulted in an unprecedented recommendation from NSABB to suspend the publication of the research.

In the United States, NSABB requested the methods used to enable transmission of the H5N1 virus from mammal to mammal and the results of the experiment be redacted before the articles were released to the public. NSABB feared that the exact details of the scientific methods used and the mutations involved could pose a significant biosafety risk. In the Netherlands, the Dutch government delayed the release of the paper by Dr Fouchier as authorities believed the work may have violated export control rules, related to dual use, and government-regulated technology [26]. The NSABB also recommended a 60-day moratorium on all similar research while the potential risks were assessed. Leading influenza scientists around the world agreed to the moratorium. The World Health Organization (WHO) did not agree with the recommendation of NSABB to redact information in the articles before publication. They believed that widely sharing the results of the articles would provide significant public health and scientific value [27]; however, the WHO did agree to the moratorium until the risks could be further evaluated [28].

Science and *Nature* both agreed to the NSABB recommendation to redact the information, with the understanding the information would be made available to researchers and organizations with a legitimate reason for access, including governments of countries with endemic H5N1, influenza research institutions around the world and pharmaceutical companies. The research groups agreed to these conditions, although members of the groups, along with others in the broader scientific community, viewed the proposal as censorship [25]. The WHO did not support making the information available only to certain groups citing that it would be too difficult and time-consuming to develop a mechanism for distribution and to determine those that would have access to the redacted information [27].

The recommendation to postpone publication and redact details of the methods used to mutate the virus to enable aerosol transmission in mammals sparked fierce debate among the scientific community, government agencies, non-governmental organizations, and biosafety specialists who all presented valid data to support the benefits and risks of such research. In the process of evaluating the potential risks and benefits of these research articles, it became clear that it is vital to assess research with DURC potential before the initiation of the project as well as before the results are submitted for publication [23]. It is interesting to note that in March 2012, the US DHHS reconvened the NSABB to review the revised manuscripts of Kawaoka and Fouchier regarding the transmissibility of A/H5N1 influenza virus in ferrets. After reviewing, NSABB decided that the revised manuscripts should be communicated in full [29].

The H5N1 dilemma highlighted a challenge for the scientific community and regulatory agencies at the intersection of two important concepts: scientific discovery and societal responsibility. Scientists must look at their research from two different points of view: adherence to accepted standards of scientific practice in both conducting and reporting research and the social consequences of applying research findings. This idea raises two important questions:

1. How much of a researcher's methods must be accessible in order to assess the integrity of the findings?
2. What is the scientist's responsibility to society at large when the research has dual use applications [30]?

The intervention of national and international government parties into the publication process reinvigorated the debate over the role of government in regulating science and the relationships between science and society. In the past, science was mostly investigations of testable hypotheses to include practical applications of new knowledge. Today, scientists often find themselves subject (at least in part) to the interests of a myriad of stakeholders – both individuals and of the commons – who see scientific results as part of a larger goal whether it be profit, funding, patients seeking a cure, politicians seeking votes or the global community seeking assurance that the science is safe. As a result, many biosafety, scientific, and government groups are demanding accountability from the scientific community. Determining the level of regulation that meets the needs of the majority of stakeholders is a balancing act that will require ongoing and open negotiation among all interested parties [30]. It has been a challenge for the government to respond to the H5N1 influenza controversy as they balance increasing public pressure for greater government oversight of scientific research with the need to support researchers in preparing for another pandemic.

2012 US policy for oversight of life sciences DURC

As a direct outcome of the H5N1 experiments, the NIH Office of Science Policy (OSP) published in March 2012, the *U.S. Policy for Oversight of Life Sciences Dual Use Research of Concern.* The policy's purpose was to promote transparency, awareness, and accountability from the inception of research through the publication of research results to protect public health. The dual use policy was not intended to restrict science but rather to support public health and national security while minimizing the risk of misuse of the technologies, by-products, information, and knowledge resulting from the research [24].

The 2012 policy was designed "to establish regular review of US Government funded or conducted research with certain high consequence pathogens and toxins for its potential DURC" [24]. The policy identified a well-defined subset of 15 of the higher-risk pathogens and toxins on the CDC BSAT list and required funding agencies to identify all federally funded research involving these agents within 60 days. Within 90 days, the agencies were required to report all instances of research involving the 15 agents that could be considered dual use. Furthermore, the funding agency, institution and lead scientists of studies found to have dual use risks were required to create risk mitigation plans. Plans could include modification of scientific methodology, relocating research to a more secure laboratory, and altering the communication of the research to the public and scientific community [24].

Previously, research executed at the NIH and CDC was reviewed for potential dual use. The 2012 policy applied this same standard to all research funded by the NIH or CDC. Dual use evaluations were now to be performed for all current and future studies for the 15 Tier I BSAT pathogens [24].

As previously discussed, the March 2012, DURC policy outlined a process for routine federal review of life science projects to identify DURC, assess the research for possible risks and benefits and methods to mitigate the risks. When the policy was promulgated in 2012, roles and responsibilities were not well defined by the USG in the document. It was assumed by many that the IBC would take the lead on the issues as they relate to DURC research because the committee has recognized knowledge of the research being conducted by investigators and because of its members' many years of experience evaluating research involving rDNA, performing risk assessments, and assigning bio-containment levels.

The policy for institutional DURC oversight

Following the 2012 DURC policy, the *United States Government Policy for Institutional Oversight of Life Sciences Dual Use Research of Concern* was officially proposed on February 21, 2013 by the OSP; this policy was published in September 2014 and became effective in September 2015 [31]. The 2014 Oversight Policy and the March 2012 DURC Policy are harmonized to accentuate a culture of responsibility to maintain the integrity of science and prevent its misuse. The 2012 policy describes the responsibilities of the federal agencies while the 2014 policy formally describes the roles and responsibilities of institutions and principal investigators conducting research that meets the criteria in the 2012 DURC policy. Of particular significance, 2014 policy reassigns the burden back to the Principal Investigator to initially identify their project as DURC. Further, the institution is required to establish an Institutional Review Entity (IRE). The IRE can be a committee established to review dual use research, an extant committee such as an IBC with the addition of *ad hoc* members to meet the established requirements of the committee, or an externally administered committee to review dual use research [31]. The IRE must be set up for the sole purpose of conducting review of research for dual use potential. The IRE must meet several criteria including that the group must be composed of at least five members and empowered by the institution to ensure it can execute the relevant requirements of the Policy for DURC Oversight. The members must have knowledge to assess the dual use potential of the range of relevant life science research conducted at a research facility. In addition, the committee must have individuals with knowledge of relevant US Government policies and understanding of risk assessment. The panel must also be conversant in risk management, biosafety, and biosecurity. The IRE is responsible for communications with the US Government funding agencies regarding mitigation plans and will have ongoing oversight of the project [31]. Not surprisingly, the requirements for the IRE align with the established mandates for an IBC. In most circumstances and at the majority of establishments, the IBC is the logical choice for the review and oversight of DURC at the institutional level. However, putting the burden of DURC review on the IBC increases the workload and regulatory burden on an already overtaxed committee that is not compensated for their time and effort. Some institutions may decide to set up a new committee or if they do not have in-house expertise on their IBC they may be forced to use an externally administered committee, such as an IBC, for the purpose of reviewing DURC protocols.

October 2014, US Government announces moratorium on gain of function research

In October 2014, the White House Office of Science and Technology Policy (OSTP) and DHHS made the announcement that they were going to review policies covering gain-of-function research projects. As noted earlier, this type research by Drs Ron Fouchier and Yoshihiro Kawaoka made world headlines in 2012 causing a moratorium on influenza research that lasted a year. Now, in the face of global threats like influenza, severe acute respiratory syndrome – Coronavirus (SARS-CoV), and Middle East respiratory syndrome – Coronavirus (MERS-CoV), which have killed many in Canada, the Middle East, and Asia, the US Government is, as of the time of writing, instituting a halt to gain-of-function funding for experiments involving these three viruses [32].

 NIH previously funded these projects at institutions in order to learn how the virus spreads, and enable scientists to assess the potential for a possible pandemic. NIH Director Francis Collins in a statement in October 2014 said these studies have biosafety and biosecurity risks that need further evaluation [32]. This moratorium has forced IBCs to re-evaluate existing protocols that include these organisms. If institutions don't have previous permission from the NIH to work with MERS-CoV or SARS-CoV then they will not receive federal funding to do the work [33]. If however, the PI already has one or both of the viruses, then they can continue working with the pathogen(s). Principle investigators are currently working with their NIH/NIAID program officers to determine whether the proposed research/experiments can be continued, weighing the risks and potential gains of the scientific outcomes. Official letters are being issued from the funding agency outlining experiments/mutants/passage studies that can or cannot be performed. The risks and benefits of the research are to be evaluated by NSABB, the NRC, and outside experts. The NSABB will determine how this type of research fits into the federal rules and regulations [33].

Conclusion

Due to the breakthrough research of Drs Berg, Boyer, and Cohen in the early 1970s involving rDNA, the NIH Guidelines were developed and implemented. When the NIH Guidelines were initially published, they were the only guidance in place for the oversight of rDNA technology. The development of the Guidelines was a direct result of the Asilomar Conference in 1975 where researchers and the public discussed their trepidation of this new emerging science. At that time, local IBCs were established at institutions receiving funds from the NIH for research involving rDNA. The IBC was originally tasked with reviewing rDNA protocols using the strict guidelines instigated at the Asilomar Conference. As research evolved in the human subject's field with recombinant gene therapy research, more IBC oversight was needed. Therefore, the IBC was tasked with reviewing human gene protocols for biosafety concerns and containment along with evaluation and mitigation of SAEs from human gene protocols. Currently, no human participant can be enrolled in a study without RAC, IBC, and IRB approval.

 The Congress enacted the "Antiterrorism and Effective Death Penalty Act of 1996" into law, regulating shipment of pathogens after the illicit ordering and delivering of *Y. pestis* to an individual's home was discovered that year. Before 1996 there was no oversight of BSAT. After 9/11, and the subsequent anthrax attacks, the threat of bioterrorism changed the landscape of regulations governing life science research. Post 9/11 in 2001, the US PATRIOT Act was passed to strengthen the BSAT program. When this regulation was passed, it required more oversight on the part of the BSO and IBC to review research rDNA protocols, SOPs, incidents and accidents, security and personnel reliability plans, and other documents to support the BSAT program.

In 2004, *"Biotechnology Research – An Age of Terrorism: Confronting the Dual Use Dilemma"* was published by the NRC and copies were provided to IBC Administrators, Chairs, and BSOs serving on the IBC. After discussing the book, IBCs around the country recognized that the federal oversight of life science research was changing in order to protect national security and public health. One of the recommendations from the 2004 book that came to fruition was the establishment of the NSABB within the Department of Health and Human Services to educate scientists and advise the government on the oversight of "dual use of concern research." As a direct result of the 2011 gain of function H5N1 influenza experiments by Fouchier and Kawaoka, the Federal Government published the US Policy for the Oversight of Life Science Dual Use Research of Concern in March 2012. Although the roles and responsibilities were not established in this first policy, it was widely accepted that IBCs would take on the responsibility of this new regulation. In preparation of what was to come, many IBCs implemented new institutional practices at this time to support the emerging regulations.

In March 2013, the NIH Guidelines were revised to also address synthetic nucleic acid molecules. The modification of the NIH Guidelines was instituted to ensure containment was assessed and biosafety guidelines were appropriate for synthetic material research [7]. As a consequence, this change to the guidelines increased the workload of the IBC by requiring an update of institutional forms to include reporting of synthetic nucleic science for review, additional training, and alterations to institutional process.

In September 2014, the Policy for Institutional DURC Oversight was released by the US Government. In this document, institutional roles and responsibilities were more clearly defined to review and oversee DURC. Many established IBCs meet the necessary criteria outlined in the policy to review DURC. As a result, most institutions have given the review of DURC to the IBC.

Policymakers and stakeholders encounter difficult decisions in the attempt to balance the possible benefits of research against potential harm such as an accidental release, gain of function projects, or bioterrorism. This balancing act was recently demonstrated in October 2014 by the Federal Government as it imposed a moratorium on SARS-CoV, MERS-CoV, and influenza gain-of-function research projects pending further review. This moratorium followed biosafety lapses at the CDC in regards to anthrax exposure to 70 employees and the unexpected discovery of 16 vials of forgotten variola virus in a storage closet at the FDA's Bethesda campus during a routine inventory. It is evident by the actions of the Federal Government that regulations and oversight guidance for life science research will continue to evolve and be further expanded as science and technology advance in the future (and, as will doubtless be the case, when new mishaps occur, even without injury or risk to humans, animals, or the environment).

IBCs – voluntarily staffed by researchers and community-based experts, and already tasked with extensive responsibilities – will face additional demands as regulations and guidelines continue to evolve, and it may be the case that these professionals will be reluctant to take additional time from already hectic schedules to address an expanding IBC agenda. Perhaps the solution is a paid committee with the appropriate expertise being responsible for the oversight of the myriad regulations at research institutions.

References

[1] Committee on Research Standards and Practices to Prevent the Destructive Application of Biotechnology. Biotechnology research in an age of terrorism. Washington, DC: The National Academies Press; 2004.

[2] The Paul Berg papers recombinant DNA technologies and researchers' responsibilities. Available from: <http://profiles.nlm.nih.gov/ps/retrieve/Narrative/CD/p-nid/260> [accessed 10.10.14].

[3] Berg P, Boyer HW, Cohen SN. Available from: <http://www.chemheritage.org/discover/online-resources/chemistry-in-history/themes/pharmaceuticals/preserving-health-with-biotechnology/berg-boyer-cohen.aspx> [accessed 10.10.14].

[4] NIH Office of Science Policy Biotechnology Assessment. Available from: <http://osp.od.nih.gov/office-biotechnology-activities/biomedical-technology-assessment/hgt/rac> [accessed 11.09.14].

[5] Office of Biotechnology Activities. Available from: <http://osp.od.nih.gov/office-biotechnology-activities> [accessed 11.09.14].

[6] Petsko G. An asilomar moment. Genome Biol 2002;3(10):1014.1–1014.3.

[7] Notice pertinent to the November 2013 revisions of the NIH Guidelines for research involving recombinant of synthetic nucleic acid molecules. Available from: <http://osp.od.nih.gov/sites/default/files/NIH_Guidelines_0.pdf>; November 2013 [accessed 12.09.14].

[8] NIH Office of Science Policy Institutional Biosafety Committees Requirements for IBC's. Available from: <http://osp.od.nih.gov/office-biotechnology-activities/biosafety/institutional-biosafety-committees> [accessed 12.09.14].

[9] What is a serious adverse event? Available from: <http://www.fda.gov/Safety/MedWatch/HowToReport/ucm053087.htm> [accessed 12.09.14].

[10] Jackson RJ, Ramsay AJ, Christensen CD, Beaton S, Hall DF, Ramshaw IA. Expression of mouse interleukin-4 by a recombinant ectromelia virus suppresses cytolytic lymphocyte responses and overcomes genetic resistance to mousepox. J Virol 2001;75:1205–10.

[11] Cello J, Paul AV, Wimmer E. Chemical synthesis of poliovirus cDNA: generation of infectious virus in the absence of natural template. Science 2002;297(5583):1016–8. Available from: <http://dx.doi.org/10.1126/science.1072266>.

[12] Rosengard AM, Liu Y, Nie YZ, Jimenez R. Variola virus immune evasion design: expression of a highly efficient inhibitor of human complement. Proc Natl Acad Sci USA 2002;99:8808–13.

[13] CDC reports potential Ebola exposure in Atlanta Lab. The Washington Post. Available from: <http://www.washingtonpost.com/national/health-science/cdc-reports-potential-ebola-exposure-in-atlanta-lab/2014/12/24/f1a9f26c-8b8e-11e4-8ff4-fb93129c9c8b_story.html> [accessed March 2015].

[14] Bureau of Industry and Security. Deemed exports and fundamental research for biological items. Available from: <http://osp.od.nih.gov/sites/default/files/resources/B_Dual_Use_Educational_Module_FINAL.pdf> [accessed 13.09.14].

[15] Federal Select Agent Program. Available from: <http://www.selectagents.gov/index.html> [accessed 26.09.14].

[16] Besser R. Oversight of select agents by the Centers for Disease Control and Prevention, U.S. Department of Health and Human Services Testimony. Available from: <http://www.hhs.gov/asl/testify/2007/10/t20071004c.html> [accessed 13.09.14].

[17] Backgrounder the select agent rule. CDC media relations. Available from: <http://www.cdc.gov/media/pressrel/b021210.htm> [accessed 13.09.14].

[18] Uniting and Strengthening America by Providing Appropriate Tools Required to Intercept and Obstruct Terrorism (USA PATRIOT ACT) Act of 2001. (Public Law 107-56, October 26, 2001).

[19] U.S. Congress. Public Health Security and Bioterrorism Preparedness and Response Act of 2002. P.L. 107-188. 42 U.S.C. 243, June 12. Available from: <http://tis.eh.doe.gov/ biosafety/library/PL107-188.pdf>.

[20] National select agent registry history of the select agent program. Available from: <http:// www.selectagents.gov/resources/42_cfr_73_final_rule.pdf> [accessed 25.09.14].

[21] National Research Council. Available from: <http://www.nationalacademies.org/nrc/> [accessed 25.09.14].

[22] Biotechnology Research in an Age of Terrorism Report in brief. Available from: <http:// sites.nationalacademies.org/cs/groups/pgasite/documents/webpage/pga_054638.pdf/> [accessed 25.09.14].

[23] National Science Advisory Board for Biosecurity (NSABB). Proposed framework for the oversight of dual use life sciences research. National Science Advisory Board for Biosecurity; 2007. Report number. Available from: <http://osp.od.nih.gov/sites/default/ files/resources/Framework%20for%20transmittal%20duplex%209-10-07.pdf> [accessed 26.09.14].

[24] National Institutes of Health, Office of Science Policy Dual Use Research of Concern. Office of Science Policy. Available from: <http://www.phe.gov/s3/dualuse/Documents/ us-policy-durc-032812.pdf> [accessed 25.09.14].

[25] Avian influenza and the dual-use research debate. Lancet Infect Dis 2012;12(3):167. http://dx.doi.org/10.1016/S1473-3099(12)70035-X.

[26] Enserink, M. Free to speak, Kawaoka reveals flu details while Fouchier stays mum. Sci Insid 2012. Available from: <http://news.sciencemag.org/scienceinsider/2012/04/free-to-speak-kawaoka-reveals-fl.html?ref=hp> [accessed 26.09.14].

[27] Cohen J. WHO group: H5n1 papers should be published in full. Science 2012;335(6071): 899–900. Available from: <http://dx.doi.org/10.1126/science.335.6071.899>.

[28] Cohen J, Enserink M. Science now one of two hotly debated h5n1 papers finally published. 2012. Available from: <http://news.sciencemag.org/sciencenow/2012/05/one-of-two-hotly-debated-h5n1-pa.html> [accessed 26.09.14].

[29] Vaccines and global health: ethics and policy NSABB Policy update: manuscripts on transmissibility of A/H5N1 influenza virus. Available from: <http://centerforvaccine-ethicsandpolicy.net/2012/04/07/nsabb-policy-update-manuscripts-on-transmissibility-of-ah5n1-influenza-virus> [accessed 26.09.14].

[30] Frankel M. Regulating the boundaries of dual-use research. Science 2012;336(6088): 1523–5. Available from: <http://dx.doi.org/10.1126/science.1221285>.

[31] United States Government Policy for Institutional Oversight of Life Sciences Dual Use Research of Concern. United States Government, 2014. Available from: <http://www.phe. gov/s3/dualuse/Pages/default.aspx>.

[32] Akst J. Moratorium on gain-of-function research. Scientist 2014 Avilable from: <http:// www.the-scientist.com/?articles.view/articleNo/41263/title/Moratorium-on-Gain-of-Function-Research/> [accessed 26.09.14].

[33] Greenfieldboyce, Nell NPR 2014 Scientists fight for superbug research as U.S. pauses funding. Available from: <http://www.npr.org/blogs/health/2014/10/23/358122198/scientists-fight-for-superbug-research-as-u-s-pauses-funding> [accessed 27.04.15].

Challenges with biocontainment facilities – building, maintaining and testing

Bruce Whitney

Chapter Outline

Introduction

Without adequate biocontainment facilities, laboratory and vivarium rooms, suites, and buildings, there cannot be adequate control of biohazards, particularly with respect to protection of individuals outside the laboratory and the environment. The document *Biosafety in Microbiological and Biomedical Laboratories (BMBL)* [1] states "[t]he term 'containment' is used in describing safe methods, facilities and equipment for managing infectious materials in the laboratory environment where they are being handled or maintained" [1]. Containment of biohazards is the goal of a biosafety program and combinations of specific elements in the triad of containment controls (i.e., safe methods, facilities and equipment) have been grouped into increasing levels of protection labeled Biosafety Levels (BSL), with BSL-1 appropriate for work with agents with the least risk and BSL-4 required for work with agents with the greatest risk.

This chapter will discuss laboratory and vivarium facilities used at Biosafety Level 2 and Biosafety Level 3. Facilities used at Biosafety Level 4, the highest level of containment and used for "exotic agents that pose a high individual risks of

Ensuring National Biosecurity. DOI: http://dx.doi.org/10.1016/B978-0-12-801885-9.00003-2

life-threatening disease by infectious aerosols and for which no treatment is available," and BSL-3-Ag, used for studies involving high-consequence livestock pathogens and large agricultural animals in which the room serves as the primary containment will not be discussed due to their highly specialized nature, extreme high cost and very limited numbers [1]. Similarly, greenhouses used for plant containment and containment for arthropods will not be covered. However, in all of these cases many of the same challenges discussed in this chapter also apply to these facilities.

There are a number of issues that owners must be aware of and handle during the lifetime of biocontainment facilities. While many of these issues are not specific to biocontainment facilities they may require more frequent actions (e.g., inspections and preventative maintenance), greater expertise (facility design, operation, and preventative maintenance), and staffing. All of this, of course, translates into a need for greater financial support.

Regulatory requirements and design guides

The starting point for understanding the challenges associated with the design and operation of biocontainment facilities is to appreciate the requirements for work with biohazardous materials. This is in itself a challenge: there can be confusion about which requirements a facility must comport with even if the specifics of the research (specific agent and type of activities) are known, which is often not the case for new facilities.

Currently, much work with biohazardous materials in the United States is not subject to federal oversight although some important exceptions are discussed below. The only actual, specific federal requirements for biocontainment facilities articulated in biosafety requirements are found in the National Institutes of Health's (NIH) document *NIH Guidelines for Research Involving Recombinant or Synthetic Nucleic Acid Molecules (NIH Guidelines)* [2]. Moreover, these requirements only apply to institutions that are registered with the NIH Office of Biotechnology Activities (OBA). These requirements are discussed below.

The *NIH Guidelines* describe facility requirements in a number of locations [2]:

- Appendix G, Physical Containment, describes the physical requirements of four biosafety levels, BL1–BL4, for laboratory work, work with small animals that can be housed in primary containment systems, or work with cultures of viable organisms of less than 10 liters in volume.
- If the volume of culture of viable organisms is greater than 10 liters, Appendix K (Physical Containment For Large Scale Uses of Organisms Containing Recombinant or Synthetic Nucleic Acid Molecules), supersedes Appendix G. Appendix K describes four biosafety levels under 'Good Large Scale Practice': BL1-Large Scale, BL2-Large Scale, and BL3-Large Scale. There is not a BL4-Large Scale biosafety level although the NIH will establish this on an individual basis if required.
- Appendix Q describes four biosafety levels (BL1-N–BL4-N) to be used for work with animals that are too large to be housed in primary containment system.
- In addition, the *NIH Guidelines* describe, in Appendix P, biosafety levels (BL1-P–BL4-P) for research involving plants.

The *BMBL*, probably the most well-known biosafety guidance document, does provide recommendations for biocontainment facilities. However, the *BMBL* is not, in itself, a regulatory document but rather a "best practice" or, as it describes itself, "both a code of practice and an authoritative reference" [1]. While not an actual regulatory document it is often used as one, both by adoption by institutions and by federal agencies. For example, while federal Select Agent Regulations ("SAR") (7 CFR Part 331, 9 CFR Part 121, and 42 CFR Part 73) [3–5] regulate possession, use, and transfer of select agents and toxins, the regulations do not actually describe biocontainment requirements; rather they require a written biosafety (9 CFR Part 121 and 42 CFR Part 73) or biocontainment (7 CFR Part 331) plan that is "commensurate with the risk of the select agent or toxin, given its intended use" [3–5]. The SAR then goes on to give references, such as the *BMBL* and the *NIH Guidelines*, that "an individual or entity should consider in developing their plan." In practice, however, the recommendations given in the *BMBL* are treated as requirements. This is best evidenced by the publication by the Federal Select Agent Program (FSAP) of biosafety inspection checklists [6] that directly incorporate sections from the *BMBL*. Other inspection checklists incorporate sections directly from the *NIH Guidelines* [6]. These checklists are used by their inspectors "[t]o ensure that entities are complying with the requirements of the select agent regulations."

The *NIH Guidelines* and the *BMBL* are similar in that they describe not only requirements for biocontainment facilities, but requirements for all aspects of biological safety (i.e., methods and equipment). In addition, these documents are similar in that the facility requirements they articulate are not described in great detail, which can lead to design and/or construction errors if those involved are not experienced with such facilities.

While the two documents described above are biosafety documents, which happen to include sections dealing with facility requirements, the following two documents are design guides, which include sections dealing with biocontainment requirements. One of these two design guides, the 2008 National Institutes of Health (NIH) *Design Requirements Manual for Biomedical Laboratories* (DRM) [7] deals specifically with biomedical research facilities. According to the NIH it is the "only detailed design requirements and guidance manual for biomedical research laboratory and animal research facilities in the U.S." [7]. While the DRM is required only for facilities owned or leased by the NIH or for facilities where the NIH funded the cost of construction, it is an extremely useful guide, when used properly (e.g. it would not be appropriate to use the heating and cooling requirements articulated in the DRM, that are specific for Maryland, for a facility in Texas). The DRM not only provides design requirements and guidance in all aspects of facility design, it also covers the pre-design and design processes. At the time of this writing, the NIH has announced that the DRM is being updated and will incorporate major changes to the 2008 version [8]. Finally, it should be noted that the NIH requires a minimum biosafety level of BSL-2 for all laboratories and vivariums subject to the DRM [9].

The second facility design guide is the USDA's *ARS Facilities Design Standards* [10]. This document "provides design policies and criteria to guide the design of ARS construction projects." This manual is unique in that it describes a fifth biosafety level, Biosafety Level 3 Agricultural (BSL-3 Ag), developed by the USDA to designate the special requirements needed for work involving high-consequence pathogens in large animals (animals of such a size that the room must function as the primary containment). Unfortunately, there was confusion by end users and designers regarding when the requirements of BSL 3 Ag were to be applied. The addition of a section on agriculture pathogen biosafety in the fifth edition of the *BMBL* in 2009 (Appendix D) provided clear guidance on when BSL-3 Ag was required [1]. Appendix D, and not the ARS Facility Design Standards Manual, now serves as the de facto BSL-3 Ag standard for the federal government, as evidenced by the use of Appendix D in the Federal Select Agent Program's BSL-3 Ag checklists [6]. As noted in the introduction, a detailed description of BSL-3 Ag requirements is beyond the scope of this chapter.

The following sections of this chapter will discuss challenges and considerations in the design, construction and operation of biocontainment facilities.

Design considerations and challenges

The greatest challenge in the design of biocontainment facilities is that both the *NIH Guidelines* and the *BMBL* lack detailed design standards. Instead they set minimum performance-based requirements, thus allowing for flexibility by IBCs and biosafety professionals in setting experiment-specific containment requirements following an appropriate risk assessment [1,2]. The disadvantage of this is that meeting the performance-based requirements is open to interpretation.

The two most important decisions, in my opinion, that are made in the entire design and construction process are selecting the Project Architect/Engineer (A/E) for the project and determining who will make the final decisions for the institution regarding end-user design requirements. An A/E without relevant experience designing facilities with the same level and type of biocontainment, with the possible exception of BSL-2 laboratories, will, almost always, lead to serious problems. These problems can include delays in the design, construction or commissioning of the laboratory as well as cost overruns, rework, to even the inability for the space to be used at the intended biosafety level.

The second decision, selecting the owner/operator's representative, is critical in that this individual will need to ensure that everyone who needs to provide input into the design of the facility is heard by the design team but at the same time ensuring that the final design meets the facility owner's overarching design and budget requirements. While it is absolutely critical that end users are involved in the programming and design phases leading to the development of the Program of Requirements (POR), a word of caution must be given against allowing investigators being given free rein to design "their labs": although buildings will be in use for 30–50 years (or more), investigators, and their research needs, come and go on a much shorter timescale. On more than one occasion I have been involved in the design of a biomedical facility where the investigator that designed a laboratory specific for his needs (desires) left the institution either before moving in to the laboratory or very soon afterwards. In several instances, renovation of the laboratory space was required before it was ever placed into use. Technology changes can also impact the requirements of laboratory design (e.g., wireless vs cable vs no computer network needs; instrument power needs [expressed per square foot], and digital imaging replacing darkroom requirements). Laboratories must then be designed with adaptability in mind as much as possible. One example of this is the use of mobile, rather than fixed casework which is increasingly common [8].

Finally, it is critical that the institution's biosafety professionals are continuously involved in the design and construction of biocontainment facilities. This is especially true during the programing and design phases where they must not be just reviewers of design and construction documents but actually involved in the design discussions from the beginning (or "a priori" if you like). As the IBC must approve the facilities prior to commencement of any activities involving biohazardous materials, it is absolutely critical that they are involved in the entire process. While this involvement may be via the institution's Biological Safety Officer, the IBC should be keep informed of, and allowed to participate in, the design and construction process. As the biosafety professionals will be called upon to provide expert advice regarding the biocontainment requirements, including interpreting federal requirements and best practices, they must have the knowledge and experience commensurate with the biosafety level of the proposed new facility. If they lack adequate knowledge and experience, which is not uncommon especially for high containment (BSL-3) laboratories, then outside biological subject matter experts must be used.

Biosafety level 2 and animal biosafety level 2 facilities

Design considerations

While laboratory requirements for BSL-2 and ABSL-2 facilities are articulated in both the *NIH Guidelines* and the *BMBL*, in general these requirements are not markedly different from those of a standard wet laboratory, for BSL-2 laboratories, and general animal facilities as described in the *Guide for the Care and Use of Laboratory Animals* (the *Guide*) [11]. In other words, a laboratory or animal facility designed for general use could, in most cases, be acceptable for use as a BSL-2 or ABSL-2 laboratory. There are, however, several important differences:

- *Biological Safety Cabinet (BSC)*: The impact of the requirement for a BSC, required for most research work at BSL-2, must be considered during the design stage. Proper location of the BSC, with respect to locations that could cause airflow fluctuations, is a commonly overlooked design element in BSL-2 laboratories. The NIH DRM proves a detailed discussion on the proper placement of BSCs within the laboratory [7]. In addition, while most BSCs used in BSL-2 laboratories do not have any special ventilation requirements, some BSCs may be connected to the laboratory exhaust system by way of a thimble connection or they may be hard-ducted. While both of these arrangements have specific design requirements, a hard-ducted BSC has a greater impact on the design and operation of the laboratory ventilation system. In addition, the hard-ducted BSC is a much more expensive option both in construction costs and in the increased operating costs due to the large amount of conditioned air removed from the laboratory.
- *Autoclave*: Appendix G-II-B-4-f of the *NIH Guidelines* requires that an autoclave is available for decontamination of waste. It does not state where the autoclave should be located. However, it can be inferred from Appendix G-II-B-2-a, which describes container

requirements for "contaminated materials that are to be decontaminated at a site away from the laboratory," that an autoclave is not required to be actually in the laboratory [2]. The *BMBL* states that a method for decontaminating all laboratory wastes should be available in the facility but in section A.8.b speaks to the container requirements for decontaminating material outside the immediate laboratory as well as when decontamination will take place outside the facility [1].

- *Handwashing sink*: While both the *NIH Guidelines* and the *BMBL* require a handwashing sink, the *BMBL* specifies that the sink "should be located near the exit door" [1,2].
- *Self-closing door*: The *BMBL*, but not the *NIH Guidelines*, articulates a requirement for self-closing doors [1,2].
- *Laboratory coat hooks*: Neither the *NIH Guidelines* nor the *BMBL* require coat hooks. However, both speak to the need to wear laboratory coats (with some possible substitutions) while in the laboratory and that these must be removed when exiting the space. The installation of coat hooks near the exit door will encourage donning of coats upon entry and the removal upon exiting.

A final design consideration for BSL-2 laboratories is whether or not to have an "open lab" design. Open laboratory designs, where the walls between laboratories are removed creating one large space, are becoming more and more common [7]. The open lab design is said to offer a number of advantages over the traditional individual laboratory space design [7]. The first advantage is that by having multiple research groups working within the same space there are increased opportunities for communication and collaboration between the different groups. A second claimed advantage of one large laboratory comparted to multiple smaller laboratories is in the cost savings that can be obtaining by reducing the number of pieces of research equipment (such as centrifuges and BSCs) and safety equipment (eyewash stations, emergency showers, hazardous chemical storage cabinets) needed, as equipment can be shared or the regulations require only one item per room (e.g., emergency shower). A significant disadvantage of the open lab concept is that when one research project uses a hazardous material or equipment, controls (biosafety levels, training and occupational health requirements, radiation use requirements) that may have applied to a few individuals working in a small laboratory may now need to be applied to the larger number of individuals working in the open lab space and supervised by different investigators. If an open lab concept is adopted, careful consideration should be given to having smaller attached rooms where work with hazards can by isolated from the general laboratory population but can still be shared, as appropriate, by those needing such facilities.

Construction considerations and challenges

BSL-2 and ABSL-2 laboratories do not present any challenges comparted to typical laboratory and animal housing space.

Operational considerations and challenges

In general, the operation and maintenance (O&M) of properly designed, constructed, and commissioned BSL-2 (ABSL-2) facilities is not notably different from that of other wet laboratories (animal facilities) found on campus. One exception is with regard to access control, required by both the *NIH Guidelines* and the *BMBL*. Methods for access by housekeeping and maintenance personnel must be developed prior to starting work and may involve having dedicated staff (trained and with the appropriate occupational health requirements in place) or having these staff, when their presence in the laboratory is required, escorted by laboratory personnel after biohazardous material has been secured and appropriate decontamination of equipment or surfaces that will be undergoing repair has been completed [1,2].

Laboratory inspections for biosafety need to be conducted on a periodic basis. Neither the *NIH Guidelines* nor the *BMBL* articulate a specific frequency for inspection of BSL-2 laboratories. The IBC should develop policies and procedures regarding inspections. These need to address how often the inspections should occur, who should conduct the inspections and what standard or criteria should be followed. In addition, the procedure should describe how deficiencies are reported and corrected.

Biosafety level 3 and animal biosafety level 3 facilities

Design considerations

The correct design of BSL-3 and ABSL-3 (i.e., high-containment) facilities looks simple enough when one reviews the requirements listed in the *NIH Guidelines* and the *BMBL*. This could not be further from the truth. While the design and construction of a small and/or uncomplicated BSL-3/ABSL-3 facility can be rather straightforward for a design team with previous experience with similar facilities, there are a number of potential issues that an unexperienced design team may not be aware of that only show up during final testing. However, even for an experienced team, challenges posed by special or unusual requirements or those caused by a never-big-enough budget can present much more challenging difficulties. As noted earlier in this chapter, choosing a Project A/E with appropriate experience is absolutely essential for a successful outcome of the project. In my experience, being cheap during the programing and design phases will cost the institution later. Prior to final selection of the A/E team, it is essential to check their references. Visit institutions (more than one if possible) that used their services and speak not only to the facilities planning and construction (FP&C) unit that managed the process but also to the IBC Chair and Biological Safety Officer (BSO), maintenance staff, and to the investigators who use the laboratory on a day-to-day basis. Not only is it important who you speak to at the reference institution, it is important whom you take to visit. The best case would be to include biosafety and maintenance staff in the visits so that each can speak to their opposite number. If everyone cannot go, make arrangements to have each specialist conduct phone interviews with their counterparts. When speaking with the FP&C unit, be aware that their criteria for success may only be that the project finished on time and budget. While this is certainly important, it is more important to determine if the facility met the biosafety requirements it was designed to meet. Ask the BSO if the facility needed to be modified prior to use to meet the standards or if the standards were changed to allow the facility to pass (it happens; a common example is that the facility does not meet the stated program requirements when the HVAC system is fault tested although it meets the minimum regulatory requirement; either the HVAC system is fixed, often at great expense or the requirement is changed to allow the less stringent requirement). See if the facility has been inspected by outside inspectors (e.g., CDC Select Agent personnel) and if so were there any facility issues identified? Ask maintenance staff if the facility is easy to maintain and whether or not there have been problems with unexpected failures. Finally, ask research staff what they like about it and what they do not like about the design and operation of the space. Ask what they would change if they could. Ask if the A/E listened to them and addressed their needs. Keep in mind that if research staff did not like something about the facility, it may be something they wanted over the objections of the A/E team or the owner/operator's representative or it may have been something that was never brought up for discussion with the design team.

When putting together the local membership of the design team, it is important to be as inclusive as possible, especially in the early stages. It is helpful to have representatives attend each meeting and the process will move forward quicker if everyone is there to provide immediate feedback and possible solutions. As stated in the beginning of this chapter, the selection of the owner/operator's representative is also critical. This individual needs to be able to provide clear directions to the FP&C and the A/E on what goes in the project and what is left out. The owner/operator's representative must have the authority to make these hard choices. It is equally important that the owner/operator's representative understands that the facility must meet all required biosafety requirements prior to use.

If the institution does not have much (or any) experience designing and building biocontainment facilities, the engagement of outside biosafety subject matter experts (SMEs) must be considered. Outside expertise can be engaged in a number of ways. At a minimum, a BSO from a peer institution could be consulted; this option may be appropriate for those institutions with some experience and needing only minimal assistance. Institutions needing greater involvement can employ a biosafety consultant or a consulting group. Their services run from periodically reviewing design and construction documents to being fully engaged as owner's agents providing not only document review but also solutions to problems that arise during the design and construction phases. An owner's agent can work with your A/E to help you ensure you do not unnecessarily over- or under-build. Finally, some owner's agents have a great deal of experience conducting commissioning of the containment facility (see the discussion regarding commissioning, below). In my experience a good biosafety SME/owner's agent/commissioning team will help ensure the project meets applicable biosafety requirements. In addition, they will help ensure the project stays on time and on budget. It will be some of the best money spent on the project. Conversely, not having a biocontainment SME/owner's agent/commissioning team, either in house or externally hired, or picking the wrong one, will with certainty, increase the cost and time of the project.

In addition to the *NIH Guidelines*, the *BMBL*, the DRM, and the *Guide*, there is an additional guidance document that should be used during the design stage: *ANSI/ASSE Z9.14 – 2014 Testing and Performance-Verification Methodologies for Ventilation Systems for Biosafety Level 3 (BSL-3) and Animal Biosafety Level 3 (ABSL-3) Facilities* (ANSI Z9.14) [12]. This is a voluntary standard that was developed by a group of stakeholders including biosafety SMEs (BSOs, consultants and commissioning agents), architects and engineers to address gaps in the performance testing of ventilation systems of high-containment facilities [12,13]. It uses "a risk assessment and performance-based approach" [12]. This document should be used in a number of ways. At a minimum, those documents listed in Section 8.2.2 of ANSI Z9.14 "Evidence of Documentation" that are or should be developed during the programming, design, construction, and commissioning phases of the project need to be accurately created and kept up-to-date. I highly recommend that a testing and performance-verification plan, as described in ANSI Z9.14, is developed with the assistance of the A/E and biosafety SME and that a contractual requirement is put in place that requires the facility to satisfactorily complete all requirements described in the plan prior to completion of the project. Doing this will ensure any corrective actions will be the responsibility of the A/E or builders and not the institution. As ANSI Z9.14 often provides multiple alternative recommended testing methodologies, the plan developed must identify a specific method for each test.

Some specific design considerations and issues include:

- The IBC and Biological Safety Officer (BSO) approve the built BSL-3/ABSL-3 facility prior to initiation of work. Specifically, the *NIH Guidelines* state in Section IV-B-2-b-(1) (ii) that the IBC "shall include" as part of their review an "assessment of the facilities" while the BSO, in Section IV-B-3-c-(1), has as one of their duties "[p]eriodic inspections to ensure that laboratory standards are rigorously followed" [1,2]. What must be kept in mind is that the IBC and BSO may require practices, equipment and facilities above and beyond that required by the *NIH Guidelines*, *BMBL* and select agent regulations. As they hold "veto power," the continuous input of the IBC and BSO is required to ensure a facility is delivered that can be used. I saw an example in which a small BSL-3 laboratory was built without involvement of the IBC/BSO until it was time to inspect the laboratory. The design and construction team, without past experience with these types of facilities, followed the *BMBL* and designed the laboratory incorrectly, specifically, the ventilation system. The cost to bring the laboratory to what the *BMBL* required was prohibitive, thus the laboratory never functioned as a BSL-3 laboratory. The cost to have built the laboratory correctly would have been a bit more than what was used but it would have been acceptable and resulted in a functional laboratory.
- Budget; the major issue in the programing and design stages is one that is no different from any construction project – staying within budget. Experienced Project A/E and biosafety SMEs can help with this but the budget must be realistic. The major expense will be the ventilation system and this is an area in which corners should not be cut no matter how tempting.
- Are the A/E's assumptions regarding meeting performance-based biosafety requirements valid? Are they even known or understood by the institution's FP&C and biosafety groups? It is not uncommon to find, at the end of the project, that the design group and the user group (i.e., biosafety oversight group) had different assumptions on requirements or how to

meet them. Are biosafety requirements even articulated in your institution's FP&C formal design guide, if one even exists? Many institutions will only state in their design guide and later in the POR that the design must meet the requirements in the *BMBL*. As noted earlier, due to the way the *NIH Guidelines* and the *BMBL* are written, meeting biosafety facility requirements is open for interpretation. This ambiguity may, at the end of a project, lead to additional work in order for the facility to obtain required biosafety approval by the BSO, IBC and/or federal select agent inspectors. In most cases these will be treated contractually as a change order requested by the institution, not a failure by the A/E to meet appropriate biosafety requirements. Thus, the cost for these changes will be the responsibility of the institution. There are a number of steps that can be taken to mitigate this common problem. First, have a biosafety SME experienced in similar construction projects involved from the beginning. Second, have clearly written and detailed biosafety requirements and assumptions as part of the POR. A useful method to identify deviations from established biosafety requirements and design guides is to write into the contract with the A/E that any deviations from the documents, that you list, must be approved in writing using a request for variance form. For example, an excellent practice is to use the NIH's DRM as the basis for comparison, fully understanding that many of the requirements in the DRM may not or should not (e.g., snow load and heating and cooling considerations for a facility deep in the heart of Texas). Another document that can be used, although as noted above not expressly written as a design document, is the ANSI Z9.14 standard. It can be used to develop a mutually agreed-upon document that articulates test requirements that the facility must meet upon completion
- The method(s) for routine and spill-response room decontamination must be determined prior to finalization of the design phase and the design must be the chosen method. Will the completed facility allow for decontamination of one room while other rooms are kept in operation? Independent decontamination of rooms within a high-containment suite will lessen the impact on other researchers but it comes at an increased cost.
- Security. Security requirements are not articulated in the *NIH Guidelines* other than to say that the "Principal Investigator controls access to the laboratory" [2]. The fifth edition of the *BMBL* added an entire section (VI) on Biosecurity [1]. As the *BMBL* notes, "[e]xcluding the Select Agent regulations, there is no current federal requirement for the development of a biosecurity program" [1]. However, that exception, the requirement under the Select Agent regulations, does necessitate some significant physical security prerequisites for locations where select agents will be stored or used [3–5]. The select agent regulations require the development and implementation of a written security plan [3–5]. The physical security system in only part of the security plan, however, choices (good and bad) made during the design phase can have an enormous impact on the ease or even the ability to conduct select agent research in a facility. A discussion of specific security concerns is beyond the scope of this document however, the federal select agent program has published guidance available to assist in the development of the required security plan as well as a security checklist [6,14].

Even if a facility is not expected to be used for select agent work at the time of design, it may be a good idea to install some of the security devices (or components thereof) at the time of construction in order to allow for future flexibility. For example, even though it is not needed at the time of design, access control to individual doors within the biocontainment suite (i.e., independent, free-standing proximity card/PIN locks or card readers wired to a door security system) may be much cheaper if installed during construction rather than after all the door frames and walls are in place.

If the laboratory is to be used for a select agent, an additional consideration is whether or not Tier 1 agents will be used; work with Tier 1 agents requires a much more elaborate (and expensive) security configuration.

- Value engineering (VE) is a process that while very important in the design process is also, from my experience, the one step that can have the largest negative impact on the final biosafety configuration of a BSL-3/ABSL-3 facility. As described by the DRM, "VE involves an organized effort to analyze alternative approaches for provision of systems, equipment, facilities, services, and supplies for the purpose of achieving the essential functions at the lowest life cycle cost consistent with required performance, reliability, quality, and safety" [8]. In theory, everyone can agree that this is a noble activity. However, where VE can come back to haunt IBCs and BSOs is when the A/E or FP&C staff have said a VE identified change will not impact the ability of the facility to operate safely when in fact it will (say during a loss of primary exhaust). A situation I have seen on a number of occasions is something is removed after VE but the change was never communicated to the biosafety members of the design team and was only found at a later date.

Commissioning considerations

Commissioning (Cx) is a process and not just a test conducted at the completion of construction. Numerous Cx activities should occur at each phase of the project and a number of groups need to be involved. It is a process to ensure all systems are "installed properly and perform according to design; cost effective; meet the users' needs; adequately documented and well understood by operators" [7]. Cx will ensure proper documentation of all systems is in place (for subsequent use by maintenance personnel as well as federal select agent inspectors if the facility will be used for such work) and appropriate training of staff to support the facility.

To obtain the maximum benefit from the Cx process, keep in mind the following:

- Cx of the BSL/ABSL-3 facility must be implemented to meet a higher standard than that of a biomedical facility without high-containment laboratories. By "higher standard" I mean that each and every device in the facility must be tested, and not a percentage as may be done for a typical building, and that the testing must be both individually and holistically.
- While Cx of typical building projects may be conducted by the construction company either directly or as one of their subcontractors, this is, in my opinion, a very poor arrangement and is rather like the fox guarding the henhouse. Unfortunately, there may be institutional resistance to having an independent third-party Cx agent hired by and reporting to the institution. This resistance may be caused by a number of reasons, some due to this not being the standard process and some due to the fear that it will cost more. Having been involved with both arrangements – construction-company-associated Cx agents and independent, owner-hired and reporting Cx agents, I can say without reservation that independent Cx agents were associated with a better outcome.
- Selection of an independent Cx agent should be as rigorous as the process described above for selection of the A/E including visiting with prior customers.

Construction considerations and challenges

Construction issues can be kept to a minimum if the A/E and Cx agent have experience in similar types of high-containment facilities. They should be able to identify problems before they become an issue. However, some things to keep in mind include:

- Someone should take photographs of the ductwork and materials running through the walls. Both wide-angle and close-up images should be taken. It is important that each image can somehow be identified in the future (i.e., as part of the file name). In my experience these photographs will be needed sometime in the future.

• Keep in mind President Ronald Reagan's adage "Doveryai, no proveryai" (Trust, but verify). While the construction manager and the FP&C representative will be involved on a day-to-day basis with the ongoing construction, an individual knowledgeable in the biosafety requirements must be involved in the oversight of the construction process. During certain times they will need to be on-site daily; at other times they may only need to visit once or twice a week. Primarily, they need to be able to verify that the materials being used and the equipment being installed are as specified in the construction documents. I have seen examples where material other than the specified material was used (or almost used). There are a number of reasons this could happen; most arising from the lack of (biosafety) knowledge on the part of the subcontractor doing the work. For example, although white caulking was specified, a subcontractor started to use clear caulking as he believed it improved the appearance of the facility, not knowing that white caulk was specified because it was much easier to be seen by internal and external inspectors who needed to verify proper surface caulking. A secondary reason for being on the job site on a frequent basis is that workers are more likely to ask for advice or identify issues when it is a familiar face that stopped to talk to them.

Operational considerations and challenges

There are a number of operational details that need to be kept in mind during the operational lifespan of a BSL-3/ABSL-3 facility. This section will discuss a number of topics that I have found to be important, both when setting up a new facility and when reviewing the operations of an operational facility.

Running a BSL-3/ABSL-3 facility is expensive. This point must be made clear to senior management. Even if salaries of support individuals are not counted, the cost associated with ongoing operations and maintenance (O&M) can add up. For example, in 2015 an institution reported net annual operating costs per square foot at a mixed use biomedical research facility as $88 for BSL-3 (non-select agent program) space, $129 for BSL-3 (select agent program) space, $129 for ABSL-3 (non-select agent program) space, and $169 for ABSL-3 (select agent program) space [15]. For comparison, their cost for general wet lab space was approximately $70 per square foot [15]. While O&M costs need to be continuously reviewed to see whether cost savings can be found, it is important that needs drive costs and not the other way away. It is clear when reading recent news stories of incidents involving BSL-3/ABSL-3 with an experienced eye, that a common root cause in many is the lack of financial support (of positions, maintenance or of the original construction). It is especially easy for institutions to fail to support ongoing maintenance.

Even before construction is completed, a number of institutional groups must work together to ensure the safe and efficient operation of the facility in such a manner that any impact on the research team is minimized. The primary groups that must work together are the biosafety team, the operations and maintenance (O&M) team and of course the research team. How this is accomplished and the amount of resources that will be required (money and staffing) will depend on the size and complexity of the facility. In the case of a large facility, representatives of each group may be part of a formal BSL-3 Operations Committee (OC). In fact, it may be appropriate that members of all three teams are part of the biocontainment facility. At the other extreme would be a small, single-room BSL-3 facility in which only three people were needed to support its function. In this case a committee would not likely be needed. A common variation of one large BSL-3 OC is to have two committees, one that coordinates between the biosafety group and the research group and one that coordinates between the biosafety group and the O&M group. While the makeup and specific functions of the BSL-3 OC must be customized to fit the local situation and most likely evolve over time, its core function of communication and coordination among different groups must be maintained. There is no regulatory requirement for a BSL-3 OC; it is just a useful tool especially for larger operations. Some examples of BSL-3 OC activities include the development of new, and review of existing, standard operating procedures (SOPs); reviewing routine and unexpected operational data (e.g., air pressure readings); scheduling of facility maintenance; coordination of and participation in the annual laboratory testing and reverification; investigations of unexpected mechanical events; and medium- and long-term planning of modifications to the laboratory or equipment. Senior management may wish to be informed of BSL-3 OC activities either by receiving minutes or even by having a representative on the actual committee.

Whether an institution decides to have some type of BSL-3 OC or not, there must be an individual identified as the final authority. While their decisions must involve the input of all stakeholders, particularly from the research groups, and whenever possible be a decision by consensus, there must be a "buck stops here person" at least for those activities that could impact biosafety in any way. This should be the Responsible Official for select agent facilities and the BSO for non-select agent facilities. In order to be effective, this individual must have the backing of senior management.

Earlier, it was stated that one of the functions of the BSL-3 OC could be the creation and review of SOPs. Accurate and comprehensive SOPs that are actually used are essential for the proper operation of all aspects of a BSL-3/ABSL-3 facility. In fact, just the act of writing a SOP serves to force a review of the actual procedure and can expose gaps and overlaps in how an activity is performed. Ideally, all SOPs needed for the operation of the high-containment facility were drafted, with the assistance of biosafety, animal care, and research staff, as part of the design and Cx activity when the facility is built but, as is often the case, they may not have been completed until well afterwards. In the case of an existing facility, the SOPs need to be in place prior to carrying out the annual testing as described in ANSI Z9.14 [12]. Poor SOPs can be worse than useless; if the instructions are not clear or actually wrong, they may direct someone to perform an unsafe activity that could be putting themselves and others at harm. ANSI Z9.14 provides a list of information that should be included in an SOP [12].

As noted previously, the *BMBL* states "[t]he BSL-3 facility design, operational parameters, and procedures must be verified and documented prior to operations. Facilities must be re-verified and documented at least annually" [1]. This requirement is, in my experience, the most often misunderstood and incorrectly implemented section of the BSL-3 requirements. Facilities that are registered to use select agents, and thus inspected by federal select agent inspectors, are now aware of how this requirement must be implemented. I would say that until about 2007, many of the registered facilities did not properly conduct this annual reverification. Even today, I would not be surprised if a large percent of non-select agent registered facilities are failing to correctly perform this activity. An incomplete understanding of how to conduct this activity was one of the driving factors for the development of the ANSI Z9.14 standard that describes a process for reverification of the ventilation systems of BSL-3/ABSL-3 [12,13]. An added benefit of this standard is that it can help upper management (and IBC members), who typically have a very limited understanding of biosafety regulations and metrics, to make an educated assessment of the state of their facilities (and biosafety program!). While the ANSI Z9.14 standard, if followed, will ensure a robust reverification of the ventilation system has been performed, it only addresses the ventilation systems of the facility. In order to completely fulfill the reverification requirement a complete reverification plan must be developed that addresses each item identified in the *BMBL*. The development of a complete plan is a straightforward (if ANSI Z9.14 is used and a model) if fairly large undertaking. For newly built facilities, the complete reverification plan should be a deliverable of the Cx process.

One area regarding the support of operational BSL-3/ABSL-3 facilities that is often found to be a cause for concern is the inability of O&M staff to provide the needed level of service that these facilities require; this is especially true for older, existing facilities but I have seen this with newer or even newly built facilities. The root cause of this deficiency most often stems, not from a lack of dedication by frontline staff, in my opinion, but rather a lack of leadership or even awareness of the issues involved on the part of senior management. Too often, it has been my experience that institutional leadership responsible for facilities did not fully understand that high-containment facilities cannot be treated as just another building. The best and most experienced O&M staff will not be able to properly support BSL-3/ABSL-3 facilities without an appropriate level of knowledge of biosafety. If O&M staff work very closely with biosafety staff, and this is highly recommended, their level of biosafety knowledge can be lower than that needed if they work independently. The model I have found most useful is to have an O&M member as a part of the biosafety team (a high-containment maintenance manager in my case) who is responsible for ensuring O&M activities are carried out as needed in a manner that protects both the research staff but also the O&M staff. This individual can be given clear expectations, appropriately trained (internally and externally) and then held accountable by the biosafety group.

As a parting note, I want to emphasize the fact that designing and building a new BSL-3/ABSL-3 facility is a challenge but this challenge can be met. The institutional challenges faced when managing an operational BSL-3/ABSL-3 are even greater. Not only is the institution needed to provide this support on an ongoing basis, failure to do so can lead to illness or even death. The common theme to meeting these tasks is having well-trained individuals that have the backing and full support of upper management. Most especially, they need to have the unwavering (and visible) support from upper management when unpopular choices with the A/E, FP&C, research groups, and budget decision makers must be made.

References

[1] Wilson DE, Chosewood LC, editors. Biosafety in Microbiological and Biomedical Laboratories (BMBL) (5th ed.). : U.S. Department of Health and Human Services; 2009. HHS Publication No. (CDC) 21-1112. Available from: http://www.cdc.gov/biosafety/publications/bmbl5/.

[2] Department of Health and Human Services, National Institutes of Health. NIH Guidelines for Research Involving Recombinant or Synthetic Nucleic Acid Molecules; 2013.

[3] 7 CFR Part 331 Possession, Use, and Transfer of Select Agents and Toxins.

[4] 9 CFR Part 121 Possession, Use, and Transfer of Select Agents and Toxins.

[5] 42 CFR Part 73 Possession, Use, and Transfer of Select Agents and Toxins.

[6] Federal Select Agent Program. Inspection Checklists [cited 2015 May 22]. Available from: <http://www.selectagents.gov/checklists.html>.

[7] United States Department of Health and Human Services, National Institutes of Health. Design Requirements Manual. In: Resources DoT, editor, 2008.

[8] NIH Design Requirements Manual 2015 Update. Design Requiements Manual News to Use 2015;1(61).

[9] Biomedical Laboratory Design Requirements. Design Requiements Manual News to Use 2012;1(1).

[10] Agriculture USDo. USDA ARS Facilities Design Standards.

[11] National Research Council (NRC). Committee for the Update of the Guide for the Care and Use of Laboratory Animals., Institute for Laboratory Animal Research (U.S.), National Academies Press (U.S.). Guide for the care and use of laboratory animals. Washington, DC: National Academies Press; 2011.

[12] ANSI/ASSE Z9.14 – 2014 Testing and Performance-Verification Methodologies for Ventilation Systems for Biosafety Level 3 (BSL-3) and Animal Biosafety Level 3 (ABSL-3) Facilities. Des Plaines, Illinois: American Society of Safety Engineers; 2014.

[13] Memarzadeh F, DiBerardinis L. Standard ANSI Z9.14: testing and performance verification methodologies for ventilation systems for Biological Safety Level 3 (BSL-3) and Animal Biological Safety Level 3 (ABSL-3) facilities. J Chem Health Saf 2011;10(02):11–21.

[14] Security Guidance for Select Agent or Toxin Facilities. In: Toxins CfDCaPCDoSAa, Program AaPHISAASA, editors.

[15] Stahl NZ. Operating Cost Benchmarks for Biomedical Research Facilities [Report]. Tradeline; 2015 [cited June 10, 2015]. Available from: <https://www.tradelineinc.com/reports/2015-2/granular-benchmarks-lead-lower-operating-costs-high-containment-facilities>.

Challenges faced by senior administration of academic institutions

Bruce Whitney

Chapter Outline

Introduction

The term "biosafety" can be defined in a number of ways. For example, the World Health Organization defines laboratory biosafety as "the containment principles, technologies and practices that are implemented to prevent the unintentional exposure to biological agents and toxins, or their accidental release" [1], while the US Department of Health and Human Services (HHS) defines biosafety as "the discipline addressing the safe handling and containment of infectious microorganisms and hazardous biological materials" [2]. Biosafety can also be defined by the regulations, rules, requirements, and best practices that an institution must, or that they have said they will, follow. As will be discussed later in this chapter, these "biosafety requirements" define what needs to be done by an institution, but aside from a few exceptions, they do not articulate who should actually perform the duties or who is responsible for ensuring that they are carried out. This is in contrast to federal regulations for research involving vertebrate animals [3] and human subjects [4], both of which require an "Institutional Official (IO)" who is responsible for ensure appropriate oversight programs and facilities (as applicable) are in place.

The upper administration in academic institutions faces a number of challenges brought about by the need to manage competing priorities and pressures to allow work with biohazardous material to proceed in a manner that ensures compliance and protects researchers, the community, and the environment while at the same time minimizing the administrative and financial burdens placed upon the scientific investigators. In addition to the typical difficulties faced in managing any type of major program at an academic institution, local oversight of a biological safety program

Ensuring National Biosecurity. DOI: http://dx.doi.org/10.1016/B978-0-12-801885-9.00004-4

faces several difficult tasks in complying with the current regulations. Following an overview of the current regulatory landscape for work with biological hazards, the roles and responsibilities within a biological safety program will be examined and challenges specific to each will be examined. Finally, the most critical component of an institution's biological safety program and the one often overlooked – the actual management of the program itself – will be discussed. It is often the case that no one individual is responsible for overall management of a biological safety program at an academic institution. In my experience, the lack of ownership by a single individual is the most common root cause of systemic failures of biological safety programs. It is my intention that this chapter will provide a roadmap for implementation of an effective and efficient biological safety program. Research at biosafety level 4 will not be specifically addressed in this chapter due to the very limited number of such facilities and their specialized nature.

Regulatory requirements

In 1976 the first federal oversight of research involving biohazardous materials came into effect, albeit with limited scope: only institutions with research supported by the National Institutes of Health (NIH) involving recombinant DNA molecules were required to follow the requirements given in the "NIH Guidelines for Research Involving Recombinant DNA Molecules" ("NIH Guidelines," renamed in 2013 the "NIH Guidelines for Research Involving Recombinant or Synthetic Nucleic Acid Molecules" following changes widening the scope of oversight to include synthetic nucleic acid molecules in addition to recombinant DNA molecules) [5]. Oversight responsibilities articulated in the *NIH Guidelines* are both decentralized and, in many cases, redundant. They identify a number of players with responsibility for the biological safety program at an institution: The institution itself, the Institutional Biosafety Committee (IBC), a Biological Safety Officer (BSO) (only required if certain types of research are conducted, for example, work with viable organisms at a large scale, defined as more than 10 liters of culture or work at biosafety levels 3 or 4), and the Principal Investigator [PI] [5]. In addition to those players named in the *NIH Guidelines*, a number of unnamed players may be needed to carryout activities described in the *NIH Guidelines* or the *Biosafety in Microbiological and Biomedical Laboratories (BMBL)* [2], such as occupational health activities, or to support named players to perform their duties. A common example of the latter is someone that assists the IBC with its paperwork (e.g., IBC coordinator). In contrast to oversight of research involving human subjects or vertebrate animals, the *NIH Guidelines* do not require anyone to be designated as an IO, that is, the individual who, on behalf of the institution, commits to ensuring compliance with the applicable regulations. Instead, Section IV-B-1 of the *NIH Guidelines* states that the institution is "responsible for ensuring that the research is conducted in full conformity with the provisions of the *NIH Guidelines.*" In addition, the *NIH Guidelines* place the responsibility for "full" compliance on the shoulders of each PI stating in Section IV-B-7 "[o]n behalf of the institution, the Principal Investigator is responsible for full compliance with the *NIH Guidelines* in the conduct of synthetic nucleic acid molecule research" [5].

In 1996, Public Law 104-132, "The Antiterrorism and Effective Death Penalty Act of 1996," was signed into law requiring the Secretary of the US Department of Health and Human Services (HHS) to regulate biological agents that have the potential to pose a severe threat to public health and safety [6]. Regulations created by HHS in response to the requirements given in Public Law 104-132 governed the interstate and intrastate transfer of these agents, termed "Select Agents" and included the registration of facilities transferring or receiving Select Agents. Included in the regulations was the requirement for an individual to be designated as a "Responsible Facility Official." This individual would "ensure management oversight of the transfer process" and should be a "senior management official" or a "safety officer" [6].

In response to the anthrax attacks in 2001, Public Law 107-56, *Uniting and Strengthening America by Providing Appropriate Tools Required to Intercept and Obstruct Terrorism (USA PATRIOT Act) Act of 2001* was created [7]. This law regulated access to Select Agents, that is biological agents and toxins that the Federal government has "determined to have the potential to pose a severe threat to public health and safety, to animal or plant health, or to animal or plant products" [8]. In 2002, broad requirements for the oversight of the possession, use, and transfer of Select Agents by the Departments of Health and Human Services (DHHS) and Agriculture (USDA) were articulated in Title II, Enhancing Controls on Dangerous Biological Agents and Toxins, of the *Public Health Security and Bioterrorism Preparedness and Response Act of 2002* [9]. Implementing regulations were initially promulgated in 2002 by the USDA (7 CFR Part 331 and 9 CFR Part 121) and the DHHS (42 CFR Part 73). These regulations specified detailed requirements for the oversight of the possession and use of Select Agents. Not only did these regulations require institutions to have a robust safety and security program in place, the regulations also implemented oversight of institutions by the appropriate federal agency (i.e., either HHS/CDC or USDA/PHIS) [10]. This oversight included registration of facilities and review of safety, security, and incident response plans and, in addition, required inspections, again by the USDA APHIS or the DHHS CDC.

The CDC/NIH publication, *Biosafety in Microbiological and Biomedical Laboratories (BMBL)*, while not a formal regulatory document, but rather, as it states, "both a code of practice and an authoritative reference" [2], is widely regarded as quasi-regulatory in nature, as the Health and Human Services Select Agent regulations explicitly state that the *BMBL* should be considered when developing biosafety plans [11]. In practice, checklists used by the CDC and APHIS inspectors to confirm appropriable biosafety requirements are in place are, for the most part, taken word-by-word from the *BMBL* [12].

Finally, there are federal Occupational Safety and Health Act (OSHA) requirements, although not all institutions are subject to their oversight [13]. While employees of state and local governments are not subject to OSHA regulations, 22 states have their own programs ("plans"), approved by OSHA, that function, for all intents and purposes, as OSHA requirements [14]. For the majority of institutions that are subject to federal OSHA oversight (or have state plans), three sets of OSHA regulations have a bearing on their biological safety program; one, 29 CFR 1910.1030 [15], directly deals with biohazards by regulating potential exposure to infectious agents. 29 CFR 1910.1030, commonly known as OSHA's Bloodborne Pathogens (BBP) Standard, addresses occupational exposure to blood and other potentially infectious materials (OPIM) [15]. 29 CFR 1910.132 indirectly regulates occupational exposure to biohazards in that it defines requirements for personal protective equipment stating "[p]rotective equipment...shall be provided, used, and maintained in a sanitary and reliable condition wherever it is necessary by reason of hazards of processes or environment..." [16]. Lastly, section 5 (a)(1) of the OSHA Act of 1970 [17] requires employers to "furnish to each of his employees employment and a place of employment, which are free from recognized hazards that are causing or are likely to cause death or serious physical harm to his employees." This General Duty Clause may be cited by OSHA in cases where a failure to follow appropriate best practices led to an employee's death or serious physical injury.

Regardless of which regulatory requirements an institution must follow (or which they follow voluntarily), their biological safety program must have, as a fundamental objective, "the containment of potentially harmful biological agents" [2]. As noted above, containment is achieved using a combination of safe methods, facilities, and equipment [2]. A risk assessment ("Comprehensive Risk Assessment" [5] or "Biological Risk Assessment" [2]) is the first and most important step in determining the appropriate containment required for work with biohazardous materials. It is also the most critical function in an institution's biological safety program as the outcome of each risk assessment will determine not only the appropriate containment but also those regulations that must be followed.

Roles and responsibilities

Responsibility for compliance with applicable federal, state, local, and institutional regulations and policies is, of course, a shared responsibility. On one hand, the Principal Investigator (PI) is the individual with the moral, if not legal, responsibility to ensure work with biohazards is conducted safely and with full compliance to applicable federal, state, local, and institutional regulations and policies. As stated in Section IV-B-7 of the *NIH Guidelines* "[o]n behalf of the institution, the Principal Investigator is responsible for full compliance with the *NIH Guidelines*..." [5]. On the other hand, the leadership of the organization is responsible for ensuring that all necessary components required for safe and compliant work with biohazards is in place before work starts. Section IV-B-1 of the *NIH Guidelines* states that the institution is also "responsible for ensuring that the research is conducted in full conformity with the provisions of the *NIH Guidelines*" [5]. In addition, the institution is the responsible entity not only in the eyes of OHSA, as noted above [17], but also in the view of the public and funding agencies.

While everyone can agree that the PI and the institution have a direct role in ensuring safe and compliant work, it is often not clear who is actually in charge of the entire biological safety program at an institution. There are several reasons why this may be. First, the requirement and best practices that most institutions follow, the *NIH Guidelines* and the *BMBL*, do not articulate a specific requirement for an IO. The Select Agent regulations do address this gap by requiring a "Responsible Official" [11,18,19] but many institutions are not required to follow these requirements as they do not fall within the scope of the Select Agent regulations (i.e., they do not possess, use, or transfer biological select agents and toxins). A second factor contributing to the common lack of a "biosafety czar" at an institution is that a biological safety program is a hybrid program combining research compliance and laboratory safety; the most common administrative structure has these two functional areas reporting to two separate senior administrators, each with their own priorities (and budgets). The roles of these administrators as well as other actors in the oversight of biosafety at the institution are now detailed.

Institutional responsibilities. As noted above, the institution is responsible for ensuring "full conformity" with the provisions of the *NIH Guidelines* [5]. In fact, an institution must not only adhere to the specifics of the *NIH Guidelines* but also its intent [5]. While the *NIH Guidelines* articulate a number of institutional responsibilities, only some of these identify a responsible party [5]. For example, the very first institutional responsibility listed, in Section IV-B-1-a, states that the institution must "[e]stablish and implement policies that provide for the safe conduct of recombinant or synthetic nucleic acid molecule research and that ensure compliance with the *NIH Guidelines*" [5]. The employee or institutional entity responsible for actually establishing and implementing these policies (among others) is not detailed in the *NIH Guidelines*, and this ambiguity allows the institution great flexibility in developing the administrative framework that best suits their unique administrate structure and research portfolio. Unfortunately, this flexibility also allows institutions to place responsibility for their biosafety oversight program at a sometimes inappropriately low level. The consequences of doing this will be discussed later in this chapter.

Other significant responsibilities of the institution, as articulated in Section IV-B of the *NIH Guidelines*, include establishing and providing training to an IBC, appointing, as appropriate, a Biological Safety Officer, appointing to the IBC individuals with plant or animal containment if such research is conducted, and determining the necessity for health surveillance and, if needed, having such a program [5]. Each of these responsibilities must be assigned to one or more individuals at the institution. In addition to the responsibilities specifically described in Section IV-B, as well as those described in other parts of Section IV and the *NIH Guidelines* as a whole, there are a number of activities implicitly required to support the biological safety program [5]. The following is a discussion of one explicitly described responsibility – an occupational health surveillance program – and three implicit responsibilities (Institutional Official (IO), IBC support staff, biocontainment support staff) of the institution.

As discussed earlier, neither the *NIH Guidelines* nor the *BMBL* explicitly designate an individual ultimately responsible for ensuring the institution fulfills its oversight responsibilities when conducting research with biological hazards, in contrast to research involving human subjects or vertebrate animals wherein federal oversight requires an individual to be designated as the IO [3,4]. The human subject regulations define the IO as the "individual authorized to act for the institution and to assume on behalf of the institution the obligations imposed by this policy" in the human subject regulations while the Animal Welfare Act regulations define the IO as "the individual at a research facility who is authorized to legally commit on behalf of the research facility that the requirements of 9 CFR Parts 1, 2, and 3 will be met." In a sense most institutions have a *de facto* IO for their biosafety program: the individual that is authorized to appoint members to the IBC. In my experience, in the majority of institutions, the Vice President of Research (VPR) is this individual. Absent a formal role and title and clearly stated authority and responsibilities that define the IO position, weak institutional leadership for the biosafety program may result. It is my strong belief that a senior administrative official must be clearly identified as the individual responsible for ensuring all biosafety requirements (i.e., federal, state, local, and institutional) are met. I further strongly believe that this individual should have a title to clearly identify them as the individual responsible for biological oversight. As the title "RO" has a very specific and narrow scope (i.e., oversight of select agents), it should not be used. I suggest the term "Institutional Official" (IO); not only does this term capture the responsibility, many individuals at an institution may equate the duties of this IO (of the biological safety program) with those of animal welfare and/or human subjects protection programs, with which they may already be aware. The duties of this IO would include all those articulated in the *NIH Guidelines* as a responsibility of the institution, that is, as listed in Section IV-B [5]. Of course, most of the actual activities in the day-to-day running of the biosafety program would be delegated to others.

The *NIH Guidelines* only require a health surveillance program for projects involving large-scale research or research with viable organisms containing recombinant or synthetic nucleic acid molecules necessitating BL3 containment (Section IV-B-1-i) [5]. Specific occupational health surveillance requirements are articulated in Appendix G-II-C-5 for research with Risk Group 3 Influenza Viruses (i.e., BL3 Enhanced) [5]. In addition to a specific requirement for projects involving large-scale activities or those conducted at BL3, it requires institutions to determine the necessity for health surveillance for all other individual projects. The Select Agent regulations only require an occupational health program if Tier 1 select agents and toxins are included on the institution's registration (although all registered entities must have policies and procedures to deal with potential or actual exposures to select agents) [11,19]. Reflecting its importance in a biological safety program, the *BMBL* has an entire section (VII) devoted to occupational health (and immunoprophylaxis) [2].

A number of different groups may be involved in an institution's occupational health program including its healthcare provider (medical support team) and someone (typically an IBC coordinator or BSO) to ensure all individuals needing occupation health clearance have been identified and have received clearance prior to working on the project. Identifying personnel needing medical surveillance and its attendant scope, along with delivery and documentation of these services is a difficult task due in part to the sheer volume of data; it is further complicated by demands for confidentiality. Many institutions already have occupational health programs supporting research with vertebrate animals or bloodborne pathogens, so some resource sharing is possible. However, it is beyond the purview of this chapter to discuss in more detail roles and responsibilities of an occupation health program due to the complexity of regulations involved and the various ways the duties described above can be divided among units internal and external to the institution.

Although a specific requirement for an individual (often with the title of IBC Coordinator) or office is not articulated in either the *NIH Guidelines* or the *BMBL*, day-to-day support for the IBC is required to assist it in carrying out its duties. In fact, the activity of this office is absolutely critical for both the compliant and efficient function of an institution's biological safety program. Just a few of the activities this office may carry out include maintaining protocol files and records, preparing correspondence to PIs and the NIH, scheduling IBC meetings, drafting meeting minutes, and tracking expiration dates of approvals and training. In addition, some activities may go beyond those supporting the IBC, such as creating and maintaining procedures for all biosafety program functions, preparing a budget, collecting statistics (i.e., number of approvals, time for approval), and providing reports to institutional leadership. The amount and nature of this support will vary depending upon the size and complexity of the biological safety program. For example, institutions with small programs having just a few active protocols may not be able to justify even one full-time employee, and instead add these duties to an existing position (or even defer any support and make the IBC Chair perform these duties). In the latter case, the institution must ensure the Chair has been given sufficient release time to handle the workload. In my opinion, a common root cause of problems in oversight at small institutions can be directly attributed to the Chair (or a staff member with multiple responsibilities) not having enough time for the work. Invariably it will not be research protocol review and approvals that fall to the bottom of the to-do list (for obvious reasons) but rather those largely unseen, but critical, activities such as tracking protocol expiration dates, sending correspondence in a timely fashion, tracking the training status of investigators and their staff, drafting meeting minutes, and creating and maintaining written procedures.

Operation and maintenance of laboratories and other facilities used for conduct of biohazardous research may be provided by a number of individuals, even at a single institution depending on the biosafety level of the facility and whether or not animals are involved. The location of the facility, physically and administratively, may impact who "owns" a facility. Even the "owner" will vary depending on circumstances; the self-professed "owner" that assigns space on a day-to-day basis (e.g., Dean or Department Chair), or actually uses the facility (e.g., the PI) may not think of themselves as responsible for funding upkeep of the facility. From a functional standpoint the PI is nominally responsible for the location of their research and at a minimum required to ensure biosafety practices are in place and followed but this may not be true or practical in shared facilities. The IBC or the BSO is responsible for inspecting the facility to ensure all required containment is in place. Even for biosafety level 1 laboratories, costs to maintain the facility (e.g., autoclaves) can add up and are not typically funded from research grants. Costs rise significantly for biosafety level 3 facilities not only because of increased utility expenses arising from the requirement for a large amount of single-pass treated air, but also the substantial requirement to test and validate the facility on an annual basis. Costs associated with operation of BSL-3 facilities are discussed in more detail in Chapter 3. Sometimes a challenge as great as finding funds to support the operation and maintenance of the facility is finding qualified personnel to do the actual maintenance work. In addition, it can be difficult to find a BSO with the knowledge and experience to test (or supervise the testing) on an annual basis the operation of a biosafety level 3 facility. See Chapter 3 for more details regarding biocontainment facilities.

Institutional Biosafety Committee (IBC). As noted previously and in Chapter 5, only certain institutions are subject to the *NIH Guidelines* and thus are required to have an IBC. While an institution may not be required to oversee recombinant or synthetic nucleic acid molecule research by an IBC due to the fact that they are not subject to the *NIH Guidelines* (e.g., they do not receive funding from the NIH), some institutions have decided to voluntarily register their IBC with the NIH, and are therefore subject to all its requirements. By agreeing to follow the *NIH Guidelines*, these institutions demonstrate their commitment to uphold current standards and best practices for oversight of this type of research including openness and transparency, and oversight. In addition, many institutions have expanded the scope of research subject to review by their IBC to include all research involving biological hazards, not just those subject to the *NIH Guidelines*. This practice is both permitted [5] and strongly encouraged [2].

The *NIH Guidelines* set forth a number of requirements and functions of IBCs in Section IV-B-2. Chief among the IBC's responsibilities is reviewing recombinant or synthetic nucleic acid molecule research conducted at the institution [5]. Section IV-B-2 lists seven specific review areas that are required to be included in the review. Five of these review areas are specific for human gene transfer experiments and are beyond the scope of this chapter. The remaining two review areas, actually the first two listed in this section, require (i) an "independent assessment of the containment levels required by the *NIH Guidelines* for the proposed research" and (ii) "an assessment of the facilities, procedures, practices, and training and expertise of personnel" [5]. The final containment requirements, both biological and physical (with the latter composed of laboratory practices and techniques, safety equipment, and laboratory facilities), are based on a risk assessment of the research. Thus, an accurate and comprehensive risk assessment is critical and arguably the single most important factor in the establishment and continued operation of a biological safety program in a safe and compliant manner. This point cannot be emphasized enough. Although all other components as described below of a biological safety program are needed, failure to properly conduct the biological risk assessment could, if the risk is determined to be less than it actually is, lead to harm to laboratory workers, the public, or the environment. On the other hand if the risk is overestimated, a waste of resources or even an improper prohibition of work that could have been done in a safe and compliant may be the result. Furthermore, the imposition of excessively stringent and unwarranted containment conditions on research may lead to research staff ignoring all or some of the mandated biological safety practices in either the targeted activities or more generally. While the IBC has the final word for setting containment requirements for five of the six categories of experiments described in the *NIH Guidelines*, from a regulatory standpoint, there are a variety of individuals that are responsible for performing the risk assessment. For example, in addition to the IBC, the *NIH Guidelines*, in Section IV-B-7-c-(1), requires the PI to "[m]ake an initial determination of the required levels of physical and biological containment..." [5]. Furthermore, the BSO (when there is a BSO) is tasked by the *NIH Guidelines* in Section IV-B-3 to provide technical advice on biosafety and laboratory security matters, both areas core to conducting a comprehensive risk assessment. The *BMBL* states the risk assessment "is an important responsibility for directors and principal investigators" but goes on to state that this responsibility is shared with "Institutional biosafety committees (IBC), animal care and use committees, biological safety professionals, and laboratory animal veterinarians" [2]. There are many models for conduct of a comprehensive risk assessment and which model is used will depend not only on complexity of the actual research, but may be influenced by the number of protocols that require review as well as the numbers and expertise of support personal. Whatever model is used, the requirement for review by the IBC at a convened meeting, of research falling into Sections III A, B, C, D, and E must always be kept in mind. Some points to consider about this process will be discussed in the section on coordination later in this chapter.

Once the risk assessment has been completed and containment conditions have been set, the second review area, an assessment of the facilities, procedures, practices, and training and expertise of personnel can be performed. As was true for the conduct of the risk assessment, there are a number of models used to perform these assessments and these will be discussed later.

Biological Safety Officer (BSO). Of all the biosafety requirements mentioned to this point, only the *NIH Guidelines* actually articulate a requirement for a BSO, and then only if the institution engages in specific types of research (i.e., activities involving recombinant or synthetic nucleic acid molecules at BL3 or BL4 or with more than 10 liters of culture) [5]. The *BMBL* does, however, strongly suggest that institutions have a BSO to assist in performing and reviewing the risk assessment [2]. It is common for an institution to appoint the individual with biosafety responsibility as BSO even not required by the *NIH Guidelines*. In my opinion, the success of an institution's biological safety program will depend upon not only the qualifications of the individual serving as the BSO but also on the workload the institution imposes on the BSO. Some institutions will assign BSO duties to an individual with other responsibilities, often a position within their environmental health and safety (EHS) office. While a position with shared responsibilities can be appropriate for institutions with a limited number of protocols and low-risk (e.g., not BL3 or BL4) laboratories, caution must be exercised so as to not overload the BSO with too many responsibilities. It is important that the institution continually evaluate the workload of the BSO to ensure there is enough time to perform all their duties. This is especially true in institutions with limited staffing as the BSO may also be required to perform those duties described in the IBC Staff/Support Office section below. This is all the more likely in that it is quite common for the BSO to have duties besides IBC-related tasks, such as laboratory safety (overseen by EHS), and therefore to report to different senior administrators who may have conflicting needs and priorities. While the workload of the BSO is important, a much more critical factor in the success of a biological safety program is the qualification of the individual named as BSO. While the *NIH Guidelines* do not speak directly to the qualifications required of a BSO, they should be knowledgeable of the science of the research likely to be conducted at the institution and also knowledgeable and have practical, hands-on experience with methods used in the research. Without a working knowledge of procedures and equipment used in a laboratory, it will be impossible to conduct either an adequate review of the research or a laboratory inspection. Not only is this knowledge and skill required to assist with their review of research, it facilitates good communication with PIs and research staff (i.e., credibility). Finally, a BSO must also have an understanding of biosafety principles and practices consummate with the type of research at the institution.

The duties of the BSO, as explicitly required by the *NIH Guidelines*, include laboratory inspections and acting as a subject matter expert regarding laboratory security and research safety [5]. The *NIH Guidelines* do note that additional duties of the BSO can be found in the *Laboratory Safety Monograph* [20]. This document describes BSO activities such as providing training, serving as point of contact with the NIH and other organizations regarding laboratory safety, and supervising emergency decontamination activities. The *Laboratory Safety Monograph* (p. 191) bluntly states that one person could not carry out all the duties listed, and should thus supervise or otherwise manage others who carry them out [20]. The *BMBL* does not provide much independent detail on the duties of a BSO but does suggest using a BSO to assist in the risk assessment process [2].

Principal Investigator or Laboratory Director. The PI, by definition, is the individual responsible for a research project and for this reason has both the knowledge and the responsibility to ensure work specific to the project is conducted in a safe and compliant manner. Legally, the PI can be held responsible for workplace-related injuries or deaths. For example, in 2011, Patrick Harran was criminally charged with three felonies in connection with the 2008 death of Sheharbano Sangiji following an accident in his UCLA laboratory [21]. The PI has a number of key responsibilities described in the *NIH Guidelines* and the *BMBL*, with key responsibilities for "the safe operation of the laboratory" and supervising "the safety performance of the laboratory staff" [2]. Two other critical responsibilities articulated in the *NIH Guidelines* include ensuring staff are adequately trained (Section IV-B7-d-(2)) and are proficient in their safety and work practices (Section IV-B-7-e-(1)) [5]. PIs are also responsible for ensuring safety equipment is functioning correctly [2,5].

Models of coordination

Rarely is an institution's biological safety program administratively located in one office or unit. This reflects the fact that a biological safety program functionally must address two different issues: safety and compliance. Typically, work with biohazards at an institution generally starts out small in size and scope and does not actually fall under any specific regulatory framework with specific requirements. For example, the first research with biohazardous material at an institution may be with a relativity safe laboratory strain of an organism such as *Escherichia coli* BL21. This organism is readily available and is commonly a component in cloning kits sold commercially. If the institution does not receive funding from the NIH, the institution falls outside the scope of the *NIH Guidelines*. Unless there are state or local regulations that require otherwise, the institution is not required to actually have a biological safety program;

responsibility for safe work resets entirely with the PI. The institution may make the decision, either formally by way of policy or informally, to follow the best practices described in the *BMBL*. Often this is done in response to the requirement of OSHA's General Duty Clause as noted earlier [17]. Later, when the institution forms an IBC, support for this research compliance committee is often located in the research division of an institution both because the need for an IBC is often driven by a term and condition of research funding and because this division is usually already supporting a research compliance committee. It just makes sense to combine the administrative functions supporting research compliance committees into one unit. There is nothing inherently wrong in having two different administrative divisions responsible for different parts of a biological safety program, as long as there is appropriate coordination between the two groups. On the other hand it is not uncommon for all functions of the biological safety program to be managed by one administrative division, typically the EH&S. The challenge in this case is not coordination between the safety and compliance functions but rather the fact that the EH&S unit may not have the knowledge and experience (and manpower) to create and run a research compliance committee. As the research portfolio grows in volume and complexity, additional regulatory requirements may need to be met and/or other functions within the program created (e.g., biosafety level 3 laboratories, work with biohazards in animal models, additional occupational health requirements and services), typically added to one of the preexisting units (safety or compliance) based on a number of factors such as availability of funding necessary to support the added functions. I have observed cases where factors that have nothing to do with the biological safety program can exert an influence on where a new component is located. Biological safety programs are not often born fully configured but usually form by aggregation.

No matter what road an institution took to arrive at its current biological safety program, a top-down review should be conducted on a periodic basis to identify any gaps (or overlaps) that may exist. The very first item that must be addressed is the selection of one senior member of the institution responsible for oversight of research involving biological hazards at the institution, that is, the "IBC Institutional Official (IO)." More often than not, one person does not have total oversight (and responsibility) either on a formal (i.e., described in a written institutional policy or rule) or informal basis. Too often, split or shared responsibility leads to a lack of accountability. It can be a challenge to identify the "IBC IO" and to codify their authority (and responsibility) but this is a critical first step. Once the "IBC IO" has been identified, the actual program review can start.

Not only should gaps and overlaps be identified, required communication between different components of the biological safety program should be identified and reviewed. Assessments that should be included in the review include whether or not the different components clearly understand what is to be communicated, to whom, and under what circumstances. It is best practice to describe in writing mutually agreed communication procedures. It is not uncommon to find errors in communication leading to instances of noncompliance (often due to assumptions being made by one or both groups).

This review should not only focus on the regulatory aspects of the program but also on the efficiency of the program – both the cost to support the program and the time required to review and approval research. In my experience, it is not often that an institution has conducted a formal top-down review of their entire program. Reviews or audits, if conducted at all, are usually limited in scope to a specific unit or group (e.g., occupational health or IBC support). In my opinion, it would be money and time well spent to perform a detailed, top-down and holistic analysis of the biosafety program initially and then as needed when there are major changes in regulations (e.g., when the select agent regulations came into force) or research programs (e.g., addition of work at biosafety level 3). Of course, any review is only as good as those doing the review and it may be useful to contact individuals at peer institutions for general or specific help or the NIH (if subject to the *Guidelines*), or the National Select Agent Program (if subject to their regulations) for answers to specific questions.

A very useful tool to assist in this review is the NIH's Institutional Biosafety Committee Self-Assessment Tool [22]. While this tool is specific for the *NIH Guidelines* it can also be used to review oversight for biohazard research not involving recombinant or synthetic nucleic acid molecules.

In the final part of this section I will include some specific points to consider regarding organization structure, coordination and evaluation of programs for the oversight of research with biological hazards. The best way to understand the different roles in the biosafety program is to follow the life of a notional project involving biological hazards.

Before work even starts, the Principal Investigator must be made aware of the need (and process) for review and approval: some type of outreach program needs to be in place. This may be passive (information on a website for example) or it may be some type of training conducted by IBC members (rarely) or by the BSO or another member of EHS.

The PI starts the process by submitting a grant proposal externally and/or internally for review. Although review and approval is not required by the IBC at this point, it is not a bad practice for protocols to be identified at this stage, either by the PI or by other means, such as those involving the use of biohazards in order to see if this research may be safely conducted with existing facilities at the institution. As some point, clearly defined by the institution, the PI submits a protocol for review. Ideally, some type of preliminary assessment is performed by the IBC support staff in order to ensure the information provided is complete enough to allow for the IBC's formal review. Many institutions with knowledgeable BSOs allow certain protocols that do not require IBC review (i.e., nonrecombinant or Section III-F) to be reviewed and approved by the BSO. A best practice would be for the BSO to report approvals to the IBC at a convened meeting.

The most common arrangement for a biosafety program, unless noted otherwise, is that IBC members are appointed by the VPR. The IBC is administratively supported either by an individual(s) in a research compliance office that also supports the IRB and IACUC or in the environmental health and safety department (EHS); both of these arrangements are common and there are advantages and disadvantages to each. The advantages to having IBC support located in the research compliance office are that the office staff is knowledgeable of best practices and experienced in performing research compliance committee support activities such as processing protocol reviews including prereview checks and entry into tracking databases, conducting meetings and writing meeting minutes, and tracking training records. There can be economies of scale with a lower number of personnel needed in the combined office, by having staff cross-trained to handle similar activities in more than one area. An additional advantage of the support office in the research compliance office, at least in institutions with large research programs, is the ability to use the same electronic software for protocol management. Use of the same software may reduce the burden on PIs in two way; first by only needing to learn how to use one system rather than two or three separate programs but also by allowing data captured (i.e., PI information) or presented (list of approved protocols) in one compliance area to be used in another. A possible disadvantage for this arrangement is that the staff generally have no (or little) biosafety experience. The absence of biosafety staff in the office managing protocol processing may lead to issues with inspections and answering biosafety-specific questions although in practice, with good communication between offices, this is not an issue. This arrangement is very similar to that commonly found in support of the IACUC, with the committee support reporting in one office while the subject matter experts (i.e., Attending Veterinarian) in a different office (although both usually report directly or indirectly to the VPR in the animal welfare program). Although the *NIH Guidelines* state that the IBC is responsible for the assessment of laboratory facilities and may in fact conduct their own inspections, most often this task is delegated to the EHS department laboratory inspectors, either as part of an overall laboratory inspection program or as separate biosafety inspections. With the BSO located in a different office there must be some method to coordinate scheduling and reporting of laboratory inspections but these are not very difficult tasks and generally do not present any issues. The disadvantage to having the BSO and IBC support in different offices – and it can the root cause of significant problems – is that the IO (i.e., VPR) does not have any control over resources assigned by EH&S to support biosafety; that is, she has responsibility without authority. Other than having the biosafety officer and occupational health program administratively colocated in the research compliance office under the direction of the IO, an unusual but not unheard of arrangement, the best method to mitigate the risks arising from the IO's lack of authority (or control) over the BSO is to have a written agreement between the IO and the BSO's Vice President. This document should clearly articulate what duties the BSO will perform for the IO, how much of the BSO's time "belongs" to the IO and which VP is responsible for finanical support for training. What often happens if there is not an existing arrangement is that VPR must convince the Vice President overseeing the EH&S department that more support is needed. This Vice President often does not have research experience and often does not understand the need for and importance of biosafety. In a number of failures of biosafety programs that I am aware of, the root cause of the incident – a lack of biosafety officer or occupational health resources (often manpower) – have been identified by the IBC and/or the "IO" before the incident but the needed changes had not been implemented in the EH&S department, often citing a lack of funds. The IBC or IO did not have the authority to force the EH&S unit to make the changes sooner (i.e., it was not a priority for the EH&S department). If this arrangement is used, the IO must have the authority to ensure enough biosafety office resources are available.

The other common arrangement colocates the IBC support staff in the EH&S department along with the biosafety officer and occupational health program, and also has some advantages. Having both IBC and biosafety support side-by-side, usually allows for a more efficient program. A disadvantage is that, as noted above, the IBC reports to a different vice president than the EHS department. This arrangement can lead to an even more problematic case of responsibility without authority: at least when the IBC support office reports to the VPR (IO), there is authority for this component. With both IBC and biosafety support outside of the VPR control, the IO has very limited ability to ensure proper funding and other resources. An additional problem when both support areas are in the same office is that it is too easy for the EH&S Director to have one individual responsible for all IBC support and biosafety activities (and often more) which can (and often does) lead to too much work to allow for proper and complete fulfillment of her responsibilities. This failure to ensure adequate staffing to fulfill all biosafety requirements in a timely and competent manner is probably the most common root cause of issues involving biosafety that I have seen.

Similar to the biosafety office, biocontainment support, if needed (e.g., BL3) is administratively located in a department that reports to a Vice President other than the IO. This Vice President almost never has knowledge or experience of biocontainment facilities and may believe such facilities can be treated as any other building on campus (as one explained to me). In an institution with only BL2 facilities, this may be true (except for a few details) but it is certainly not true of BL3 facilities. See Chapter 3 for an in-depth discussion on management of biocontainment facilities.

In summary, in order to have a strong and successful program of oversight of research with biological hazards there must be a senior administrative official clearly designated as the leader of the program. This individual must have the authority and means to ensure a program is put in place and maintained to ensure continuing competence and adequate staffing. The VPR is the ideal individual to be named the IO for the biosafety program as she has an incentive to ensure an effect and efficient program is in place in order to support the PI's research efforts. While some may argue that having the VPR oversee this program is like having the fox guard the henhouse, it has been my experience that VPRs know that a failure in their compliance program is a risk that they cannot take as it can have a devastating impact on their entire program. The IO must ensure that critical personnel, but in particular the BSO, have the experience and training necessary to fulfill their duties competently. Unfortunately, it is not uncommon to see an individual appointed as BSO not based on their biosafety knowledge but on the fact that they are available in the EHS department and doing something that is thought to be similar (i.e., chemical hygiene officer). This is often done due to a lack of funding to hire a qualified person and coupled with not providing training, again in an effort to save money, leads to a vacuum of biosafety knowledge at the institution often not appreciated by the IO. Finally, the IO must, on a periodic basis, ensue appropriate reviews of the biosafety program are conducted. As described earlier in this chapter, the NIH OBA has a self-assessment tool that should be used for research subject to the *NIH Guidelines* and, with minor changes, can be used for all research subject to oversight by the biosafety program. However, the review should be wider in scope that just this but include a review of approval times, inspection results (i.e., are the same deficiencies coming up time after time?), attendance of members to the IBC, and the amount of training provided. I believe a report to the IO, maybe on an annual basis by the IBC Chair as well as the biosafety support offices, helps ensure the IO stays engaged.

As noted throughout this chapter, the key to success is a visible, engaged IO with the authority to carry out their duties.

References

[1] World Health Organization. Laboratory Biorisk Management Strategic Framework for Action 2012–2016. World Health Organization, 2012; WHO/HSE/2012.3. Available from: www.who.int/ihr/publications/strategic_framework/en/.

[2] Wilson DE, Chosewood LC, editors. Biosafety in microbiological and biomedical laboratories (BMBL) (5th ed.). : U.S. Department of Health and Human Services; 2009. HHS Publication No. (CDC) 21-1112. Available from: <http://www.cdc.gov/biosafety/publications/bmbl5/>.

[3] 9 CFR Parts 1-4 Animal Welfare Regulations.

[4] 45 CFR Part 46 Protection of Human Subjects.

[5] Department of Health and Human Services, National Institutes of Health. NIH Guidelines for Research Involving Recombinant or Synthetic Nucleic Acid Molecules. 2013.

[6] Public Law 104-132 The Antiterrorism and Effective Death Penalty Act of 1996. 1996.

[7] Public Law 107-56, Uniting and Strengthening America by Providing Appropriate Tools Required to Intercept and Obstruct Terrorism Act of 2001 (USA PATRIOT Act). 2001.

[8] Federal Select Agent Program – General FAQ's About Select Agents and Toxins. Available from: <http://www.selectagents.gov/faq-general.html>.

[9] U.S. Congress. Public Health Security and Bioterrorism Preparedness and Response Act of 2002. P.L. 107-188. 42 U.S.C. 243, June 12. Available from: <http://tis.eh.doe.gov/biosafety/library/PL107-188.pdf>.

[10] Federal Select Agent Program, History, Bioterrorism: A Brief History. Available from: <http://www.selectagents.gov/history.html>.

[11] 42 CFR Part 73 Possession, Use, and Transfer of Select Agents and Toxins.

[12] Federal Select Agent Program. Inspection Checklists [cited 2015 May 22]. Available from: <http://www.selectagents.gov/checklists.html>.

[13] OSHA Frequently Asked Questions – Bloodborne Pathogens. Available from: <https://www.osha.gov/OSHA_FAQs.html>.

[14] OSHA – What is an OSHA-approved State Plan. Available from: <https://www.osha.gov/dcsp/osp/index.html>.

[15] 29 CFR 1910.1030 Bloodborne Pathogens.

[16] CFR 1910.132 Personal Protective Equipment.

[17] Public Law: 91-596 Occupational Safety and Health Act of 1970.

[18] 7 CFR Part 331 Possession, Use, and Transfer of Select Agents and Toxins.

[19] 9 CFR Part 121 Possession, Use, and Transfer of Select Agents and Toxins.

[20] Department of Health and Human Services, National Institutes of Health. Laboratory Safety Monograph – A supplement to the NIH guidelines for recombinant DNA research. 1978.

[21] Hayden E. Chemist reaches agreement with prosecutors over lab death. Nat. News 2014. Available from: <http://www.nature.com/news/chemist-reaches-agreement-with-prosecutors-over-lab-death-1.15444>.

[22] Institutional Biosafety Committee Self-Assessment Tool [cited 2015 May 24, 2015]. Available from: <http://osp.od.nih.gov/sites/default/files/resources/2014%20IBC%20Self%20Assessment.pdf>.

Overcoming regulatory gaps in biological materials oversight by enhancing IBC protocol review

David Rainer and Susan Cook

Chapter Outline

Introduction

> *Science may set limits to knowledge, but should not set limits to imagination.*
>
> Bertrand Russell [1]

Scientists have boundless energy, initiative, independence, and imagination and excel at developing and changing research initiatives. So too must the safety review process evolve to help ensure the maintenance of a safe work environment.

Ensuring National Biosecurity. DOI: http://dx.doi.org/10.1016/B978-0-12-801885-9.00005-6

Although the regulatory mandate related to the role of the Institutional Biosafety Committee (IBC) has evolved over 30 years and become more expansive, institutions are self-regulating and are therefore filling gaps in non-regulated research by having IBCs conduct safety reviews beyond their federally mandated requirements.

Most interesting is what is *not* prescribed in the NIH *Guidelines* for staffing IBCs [2]. The *Guidelines* are silent on the issue of staffing of an IBC administrator, environmental health and safety (EHS) professionals, and others who may be relevant to the safety review process. However, regulatory gaps in safety review oversight are effectively being managed by including many of these professionals on IBCs and in "scope creep" of IBCs. Scope creep is having a profound impact on the types of research these committees review, the way IBCs operate and the time it takes these committees and their dedicated appointees to perform their jobs related to review. We will detail the evolving responsibilities—both those already realized and those likely to be in the future—later in this chapter.

By way of an example, North Carolina State University requires investigators, diagnostic lab directors, and course instructors to receive IBC approval prior to obtaining or using any of the biological materials listed below [3]:

- Recombinant or synthetic nucleic acid molecules including their use in animals (including arthropods) and plants
- Human and other primate-derived substances (blood, body fluids, cell lines or tissues)
- Organisms or viruses infectious to humans, animals, or plants (e.g., parasites, viruses, bacteria, fungi, prions, rickettsia) or biological materials that may contain these microorganisms
- Select Agents or Toxins (human, animal, or plant)
- Biologically active agents (e.g., toxins, venoms) that may cause disease in humans or cause significant impact if released to the environment.

Agkistrodon piscivorus, Agkistrodon contortrix, Boulengerina annulata, Naja mossambica, Naja haje annulifera, Thelotornis capensis, Micrurus fulvius. Unsure of what these are? So was the North Carolina State's Institutional Animal Care and Use Committee (IACUC) when it went on a herpetology facility inspection and asked for an up-to-date species list to better understand the potential hazard of the venomous snakes available for milking and research on the campus. When the species names were translated to common names, cottonmouth, copperhead, ringed water cobra, spitting cobra, Egyptian cobra, twig snake and coral snake, there was some significant anxiety over the presumed hazard especially since the snakes were being milked to collect venom to analyze chemical make-up and potential use for medical treatment.

But who would be tasked to more fully understand the hazards and make recommendations about risk tolerance associated with the snake research? The IBC, of course, with subject matter experts including the state zoo herpetologist and state science museum reptile curator. Not only did the IBC assess the level of risk but it also went on to assess safe snake-handling practices, facility design and containment, signage to alert emergency facility responders, security, and campus emergency response to snake bites. The IBC also initiated hospital contact to ensure local emergency rooms could respond appropriately.

After initial IBC deliberations about risk tolerance associated with the research program with these snakes the committee made a recommendation to the Principal Investigator (PI) and Vice Chancellor for Research to send back some of the snakes to their owners (some of the snakes were on loan) due to lack of antivenin and the likelihood of significant medical emergency or death resulting from a bite. The recommendations were immediately accepted and an IBC representative oversaw the packing and shipping of some specimens back to their owners (if you are envisioning snakes on a plane, you are correct).

This chapter will cover a range of subject areas related to what IBCs do. But, as is already apparent, since there are no limits to scientists' imaginations, campuses will always be challenged to assess and review new research protocols with all manner of biologically active agents and toxins. A logical place for review is the IBC because of its diverse make up and reporting structure through senior university administration. This chapter will also discuss regulatory gaps that exist in oversight of research with infectious agents and recombinant or synthetic nucleic acids.

Functions of an IBC

Functions mandated by the *NIH* Guidelines

In Chapter 1, Zelicoff defined the terms biosecurity and biosafety. "The term biosecurity refers to the protection, control of, and accountability for high-consequence biological agents and toxins and critical relevant biological materials and information within laboratories to prevent unauthorized possession, loss, theft misuse, diversion or intentional release. Biosecurity is achieved through an aggregate of practices including the education and training of laboratory personnel, security risk assessments, Biological Select Agent and Toxin (BSAT) access controls, physical security (facility) safeguards and the regulated transport of BSAT. Achieving effective biosecurity for BSAT is a shared responsibility between the Federal Government and facilities/ individuals that possess, use or transfer BSAT."

"Complementary to, but distinct from biosecurity is *biosafety* based on principles of *containment* and *risk assessment* in the laboratory." Containment includes: "the microbiological practices, safety equipment, and facility safeguards that protect laboratory workers, the environment and the public from exposure to infectious microorganisms that are handled in the laboratory," whereas risk assessment is "the process that enables the appropriate selection of microbiological practices, safety equipment, and facility safeguards that can prevent laboratory-associated infections." In short, biosecurity refers to the tools to manage *threats*, whereas biosafety procedures mitigate *risks*.

The *NIH Guidelines for Research Involving Recombinant or Synthetic Nucleic Acid Molecules (NIH Guidelines)* are constantly evolving and are extremely comprehensive for managing recombinant or synthetic nucleic acid research [4]. As stated in the scope, "The purpose of the NIH *Guidelines* is to specify the practices for construction and handling of: (i) recombinant nucleic acid molecules, including those that are chemically or otherwise modified but can base pair with naturally occurring nucleic acid molecules, and (ii) cells, organisms, and viruses containing such molecules" [4].

Also according to the *Guidelines*, Section IV-B-2-b Functions, the IBC is responsible for, among other things:

1. Reviewing recombinant or synthetic nucleic acid molecule research conducted at or sponsored by the institution, including containment levels required; assessment of the facilities, procedures, practices, training, and expertise of personnel involved;
2. Assuring conformance to Section M of the *Guidelines* for research involving the transfer of recombinant or synthetic nucleic acid molecules into one or more human research participants, and
3. Adopting emergency plans covering accidental spills and personnel contamination.

Functions not mandated by the Guidelines *but often delegated to IBCs*

The reality is that the types and complexity of research programs are increasing and disciplines like engineering, chemical engineering, mechanical and aerospace engineering, material science and other physical sciences are moving beyond the traditional boundaries of their respective research roles and looking at biologically active systems and toxins.

While the *Guidelines* provide an extremely important regulatory and health and safety framework, one should not lose sight of the fact that from the operational and regulatory perspective of a campus EHS group, safety oversight and safety reviews are required for innumerable projects not explicitly covered by the *Guidelines*. So, how can academic EHS staff piggyback on the *Guidelines* to provide practical project review? What gaps exist in current government-mandated safety reviews? And how do EHS programs support risk assessment and hazard review in a timely manner so as to manage the risks of working with biological systems and agents? A brief brainstorming session among EHS staff may identify many programs with biological systems and agents and toxins requiring review, including but not limited to the following:

- Exempt recombinant or synthetic nucleic acid research
- Non-recombinant pathogens (human, animal, plant)
- Handling human and animal specimens, tissues, and cell lines
- Wild animals/field specimen collection
- Dual Use Research of Concern (DURC)
- Gain-of-function research
- Laboratory safety audits
- Chemical safety (including engineered nanomaterials)
- Review of laboratory design (BSL-3 and Laboratory Commissioning)
- Biologically active agents (e.g., toxins and venoms) that may cause disease in humans or cause significant impact if released to the environment
- Standard Operating Procedure (SOP) review of animal handling practices, medical surveillance, waste management, and disposal
- Collaborate to facilitate review with IACUC, Institutional Review Board (IRB), and other campus safety committees.

The reality is that many organizations are using the expertise and organizational structure of their respective IBCs to support safety and health reviews of items like those listed above which are not defined within the scope of the NIH *Guidelines*. Because IBCs by nature are interdisciplinary teams, they are uniquely positioned to support safety and health reviews and serve as a safety gatekeeper for many kinds of research protocols. This can be both a blessing and a curse because the committees provide a framework and structure for conducting safety reviews recognizing that the time commitment may be onerous and academic departments may not reward participation on safety and health committees. In the long term though, the benefit of using an IBC to conduct health and safety reviews is a worthy initiative and serves to protect the institution, its employees, and students by managing risk. The physical damage to life and limb, inadvertent exposure to biological agents or toxins to persons or the environment and potential damage to reputation caused by laboratory and research program accidents should convince anyone who questions this expansion of IBC duties to pause and see the logic of supporting IBC safety reviews outside the boundaries of their traditional role.

IBC engagement and managing safety and health reviews (NC State case study)

Let's suppose that your institution is thinking about a Center for Programmable Plants (CePP) [5,6]. The Center will be looking at the whole plant to improve economically relevant traits in plants such as improving shelf life, limiting potential for damage/ bruising during shipment, and maximizing food safety. The CePP will bring together expertise in mathematical modeling and engineering with plant system biologists to understand the underlying networks that contribute to yield variability and traits most wanted by consumers.

The plant disease group will be looking at plant disease, plant biotic stress, plant–fungal interaction, transgenic plants for disease resistance, virus disease, sustainable agriculture in the developing world, and the ecology of plant biotic interactions.

The abiotic stress group will be studying drought stress, extremophile genes for engineering plant stress tolerance, systems biology, plant metabolic engineering, physiological ecology of plants, climate change, drought ecology, and the intersection of time and temperature on the control of plant response to osmotic stresses.

The nutrient stress and availability group will be evaluating systems biology, plant iron and nutrient stress, modeling of regulatory pathways, characterizing soil microorganisms and microbial processes using culture-dependent and independent methods.

Because the CePP is a major program for discovery and innovation and will bring together interdisciplinary teams and require business and academic partnerships this plant science initiative will require a world-class facility to bring the ideas to life (Figure 5.1).

Assuming that a project with this scope and breadth has the approvals of senior university administration, it is also apparent that this project (really several interlocking

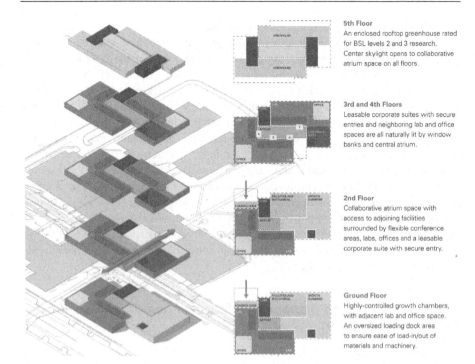

5th Floor
An enclosed rooftop greenhouse rated
for BSL levels 2 and 3 research.
Center skylight opens to collaborative
atrium space on all floors.

3rd and 4th Floors
Leasable corporate suites with secure
entries and neighboring lab and office
spaces are all naturally lit by window
banks and central atrium.

2nd Floor
Collaborative atrium space with
access to adjoining facilities
surrounded by flexible conference
areas, labs, offices and a leasable
corporate suite with secure entry.

Ground Floor
Highly-controlled growth chambers,
with adjacent lab and office space.
An oversized loading dock area
to ensure ease of load-in/out of
materials and machinery.

Figure 5.1 Proposed NC State University plant sciences facility.

projects) will require navigation through multiple complex reviews and approvals that
need to complement each other. Although only a portion of the project requires the
construction and handling of recombinant nucleic acid molecules meeting the criteria
defined for review under the *Guidelines*, many campuses would use the resources of
the IBC to begin a review. So what is a logical first step?

Since this project will require intramural and extramural funding, let's assume the
project has the needed on-campus approvals from the appropriate Deans and university executive staff to proceed. As the project concept moves forward, there needs to
be an administrative process to assure the engagement of the applicable academic and
administrative departments: everyone from the University Architect (new building
siting, design, aesthetics, budget), EHS (building safety, biosafety, risk assessment,
fire safety), Research Administration (contracts, grants, regulatory reviews), and others must be included. One of the initial reviews should be through the IBC and one of
the first questions that should be answered by IBC review is: Is the scope of the project sufficiently defined so that a conclusion may be drawn regarding safe conduct of
that research on your campus? This question, challenging in both depth and breadth,
must be evaluated both in the context of your campus's safety, regulatory and risk
management process as well as in the context of whether or not the proposed facility
is adequate in terms of design and operation to support the research program.

IBC membership

Based on *NIH Guidelines* (section IV-B-2-a-c1), the minimum number of IBC members is five. The *Guidelines* specify that the members "collectively have experience and expertise in recombinant or synthetic nucleic acid molecule technology and the capability to assess the safety of recombinant or synthetic nucleic acid molecule research and to identify any potential risk to public health or the environment" [4].

However, as IBCs move beyond their traditional scope and evaluate all manner and types of projects the expertise represented by committee members will inevitably move beyond traditional boundaries.

It is clear from the scope of the research program outlined above that it is not likely that a five-member committee could have all of the requisite expertise to review an expansive research program like the CePP, and we will discuss the additional expertise required as part of this case study. A more well-rounded committee and a committee structure more likely found in an organization with a complex research agenda may preferentially include: scientists, clinical investigators and administrators (with a mix of tenured and non-tenured faculty) that have varied technical expertise (e.g., recombinant DNA; virology, microbiology, agents infectious to humans, animals or plants; acute toxins of biological origin, risk assessment, medicine, veterinary medicine, physical security), and other expertise relevant to the type of protocols being reviewed.

Special membership requirements

Community representation
NIH *Guidelines* stipulate that at least two members of the IBC not be affiliated (non-affiliated members) with the institution, but represent the interests of the local community and general population. Community members may include officials of state or local public health or environmental protection agencies, members of other local governmental bodies, or persons active in medical, occupational health, or environmental concerns in the community. While the NIH *Guidelines* do not stipulate a particular educational background for a non-affiliated member, this person(s) must be able to understand the basic concepts of the registration(s) submitted to the committee, understand relevant biosafety and IBC review procedures at the respective institution and must have a conflict-of-interest statement on file with the institution. For example, NC State community members have expertise in plant and molecular biology and bioterrorism and emerging pathogens.

The IBC should also have *ex-officio members*: by virtue of their administrative or regulatory positions, the Biosafety Officer (BSO), the Director of EHS, and other relevant compliance personnel should be *ex-officio* members of the IBC. Based on the NIH *Guidelines* (Section IV-B-3-b) the BSO is a voting member if the institution engages in recombinant or synthetic nucleic acid molecule activities requiring BSL-3 containment and/or large-scale (>10L) recombinant or synthetic nucleic acid molecule activities. However, we believe that it always makes sense for the BSO to be a voting committee member.

Subject matter expertise

Most important if a registration is outside the area of expertise of IBC members, the
IBC Chair should always be authorized to seek counsel from one or more individuals
knowledgeable in the subject matter. This person could be someone external to the
organization as necessary. Recall that in the introduction to this chapter the IBC was
tasked to make risk management decisions related to the use of venomous snakes and
the Vice Chancellor of Research wisely asked subject matter experts from the state
zoo (herpetologist) and state science museum (reptile curator) to participate in the
review.

Research projects utilizing specialized materials or incorporating novel systems
such as the use of nanoparticles would certainly warrant bringing in outside subject
matter expertise. Another issue requiring expertise is physical security. Whether it be
your committee's recommendation that drives physical security enhancements based
on a risk assessment or regulatory entity request, deploying appropriate security systems requires a knowledge base distinct from biological and physical sciences.

In summary, as you consider the makeup of your IBC and the expertise of committee members, recognize that a diversity of expertise and disciplines in committee
makeup helps effectively carry out expected responsibilities.

Standardizing the IBC review process

Figure 5.2 provides a schematic layout for an IBC review process. The NIH *Guidelines*
outline the minimum responsibilities for the institution, IBC, BSO, PI, and subject
matter experts. As discussed elsewhere in this chapter, many institutions choose to
assign additional responsibilities to the IBC, BSO, PI, and subject matter experts.
For ease of discussion, we will take North Carolina State University (NC State) as an
example of a typical academic research institution, and will discuss the overall IBC
review process, as well as the specific responsibilities assigned to each party.

This particular institution (NC State) has chosen to assign its IBC the responsibility of reviewing all research with biohazard risks that include the following biological
materials:

1. Recombinant or synthetic nucleic acid molecules in organisms
2. Creation of transgenic plants or animals
3. Human and other primate-derived substances (blood, body fluids, cell lines, or tissues)
4. Organisms and viruses infectious to humans, animals, or plants (e.g., parasites, viruses,
   bacteria, fungi, prions, rickettsia)
5. Biologically active agents (i.e., toxins, allergens, venoms) that may cause disease or injury
   to other living organisms or cause significant impact to the environment or community.

IBC Responsibilities

The IBC provides recommendations to administration officials responsible for conduct of research at the university. These administrators have the responsibility to

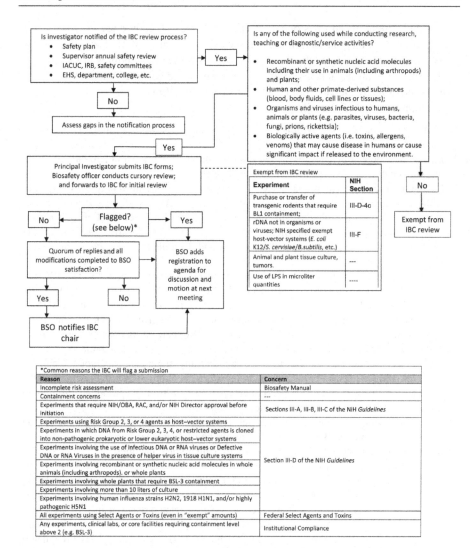

Figure 5.2 IBC protocol review flow.

facilitate carrying out the recommendations of the IBC. EHS staff usually process administrative tasks of the IBC.

IBC responsibilities extend and apply to all instructional and research projects conducted at the university, regardless of location on the property, including all rented or leased facilities. Requirements also apply if work is being done off-site. The local IBC may accept the approval of another IBC or may officially delegate an external IBC to act on its behalf.

The IBC establishes, recommends, and/or approves policies on the proper use of biohazardous agents including, but not limited to: recombinant or synthetic nucleic acid molecules, transgenic animals and plants, infectious agents, acute biological toxins, and venomous animals/poisonous plants. Policy objectives are to protect staff, research subjects, the general public, and the environment from biohazardous agents. In the unlikely event that a laboratory persists in following procedures in violation of compliance regulations and IBC policies, the Committee will recommend the imposition of sanctions by Department Heads, Deans, and/or Provost.

In summary, The IBC shall:

1. Establish and monitor policy, practices, and procedures for work involving biohazardous agents at the university.
2. Ensure that adopted policies, practices, and procedures for work with biohazardous agents meet applicable regulatory standards and guidelines.
3. Review biological research conducted at or sponsored by the university for compliance with adopted policies, regulations, and guidelines. This review shall include an independent assessment of the biological containment required, and an assessment of the facilities, training, and expertise of personnel involved in the research. The IBC shall ensure that the PI is provided with the results of the review and determination of approval in a timely manner.
4. Assess proposed containment facilities and practices for research projects. The IBC will use the biosafety levels (BSL) published by the Centers for Disease Control and Prevision (CDC), National Institutes of Health (NIH), and US Department of Agriculture (USDA) as the usual standards of containment to be set for work with a given biological agent. To the extent allowed by Federal law and regulation, the IBC may, at its discretion, increase or reduce the BSL depending on the circumstances presented by a specific project.
5. Review any findings of the BSO in investigating any significant violation of policies, practices, and procedures; participate in an investigation of any significant research-related accidents or illnesses; and recommend to university administration appropriate disciplinary action if an investigation reveals significant violations.
6. Perform such other functions as may be delegated to the IBC by the university administration.
7. In conjunction with the BSO, responsibilities of the IBC include:
   a. Review design specifications and criteria for containment facilities.
   b. Review and certification of BSL-3 laboratories.
   c. Review and final approval of policies and procedures related to select agents, including access, inventory management, laboratory protocols, and emergency response plans. Review and assessment of compliance with permit-related requirements for work with materials from USDA Animal and Plant Health Inspection Service (APHIS), Veterinary Services (VS), Plant Protection and Quarantine (PPQ), and Biotechnology Regulatory Services (BRS); and Environmental Protection Agency (EPA).

Members of the committee review research protocols. However, it is not the responsibility of the IBC to critique a PI's research program. Applications are reviewed for biological containment and biological safety concerns only.

Biological Safety Officer responsibilities

A BSO position in EHS assists the IBC and the research community in meeting the compliance requirements of the NIH *Guidelines*. The BSO is also responsible for the development, implementation, and maintenance of a comprehensive Biosafety Program. The IBC and its policies complement and support the objectives of this

Biosafety Program and the BSO is often the primary intermediary between PIs and the IBC. The specific tasks for the BSO are to:

1. Manage the Biosafety Program and support implementation of IBC policies and procedures.
2. Maintain an active IBC and recommend the appointment of the IBC Chair, Vice-Chair, and committee members to the university administration.
3. Assist laboratories in conforming to pertinent regulatory guidelines and IBC policies by providing training, facility inspections, and communication of Biosafety Program and related regulatory requirements.
4. Perform annual inspections of BSL-2 and BSL-3 laboratories for compliance with NIH *Guidelines*, the CDC *publication Biosafety in Microbiological and Biomedical Laboratories* [7], and the OSHA Bloodborne Pathogens Standard as applicable.
5. Screen research protocols submitted by PIs and make recommendations to the IBC.
6. Prepare periodic reports for institutional management regarding IBC activities and Biosafety Program status.
7. Screen protocols submitted to the IACUC for identification of biological hazards; consult with animal facility management, veterinarians, and PIs regarding appropriate containment procedures for biohazardous agents.
8. Ensure preparation of minutes of IBC meetings.
9. Monitor federal, state, and local regulatory trends, and communicate any changes to the IBC.
10. Submit required annual reports to NIH Office of Biotechnology Activities.

Director of the Environmental Health and Safety Center responsibilities

The Director of the Environmental Health and Safety Center, through the BSO and the IBC, implements a program to help ensure that research is conducted in full conformity with local, state, and federal policies and regulations. In order to fulfill this responsibility, the Director of the Environmental Health and Safety Center's tasks are to:

1. Establish and implement policies that provide for the safe conduct of research and teaching involving biohazardous agents.
2. Through the IBC and the BSO, assess compliance with the regulations and guidelines by PIs conducting research at the university.
3. Review the adequacy of resources for the dissemination of information on biohazardous agents and biosafety procedures, including training programs and workshops.
4. Review resources for medical surveillance measures to protect the health and safety of employees.
5. Represent the IBC as needed in university functions.

Principal Investigator responsibilities

The PI is defined as the faculty member or other person acting in their official capacity as a university representative who leads the research effort and is ultimately responsible for the conduct of research within the space assigned to his/her. The PI is responsible for full compliance with the policies, practices, and procedures set forth by the university. This responsibility extends to all aspects of biosafety involving all individuals who enter or work in the PI's laboratory or collaborate in carrying out the PI's research. Although the PI may choose to delegate aspects of the Biosafety Program in his/her laboratory to other laboratory personnel (laboratory directors or supervisors) or faculty, this does not absolve the PI of his/her ultimate responsibility. The PI remains accountable for all activities occurring in his/her laboratory. Documentation of training and compliance with appropriate biosafety practices and procedures is essential. The PI is responsible for assuring the appropriate safety training of employees and for correcting errors and unsafe working conditions.

As part of general responsibilities the PI shall

1. Develop and implement written laboratory-specific biosafety procedures that are consistent with the nature of current and planned research activities and make available copies of the specific biosafety procedures in each laboratory facility. The PI shall ensure that all laboratory personnel, including other faculty members, understand and comply with these laboratory-specific biosafety procedures.
2. Delay initiation of research until the research protocol has been approved by the IBC. Ensure that all laboratory personnel, maintenance personnel, and visitors who may be exposed to any biohazardous agents are informed in advance of their potential risk and of the behavior required to minimize that risk. It is essential that everyone who may have potential exposure to biohazardous agents be informed of such hazards and appropriate safety practices before entering or working around or with such hazards.
3. Ensure that all maintenance work in, on or around contaminated equipment is conducted only after that equipment is thoroughly decontaminated by the laboratory staff or PI.
4. Ensure that research materials are properly decontaminated before disposal and that all employees are familiar with the appropriate methods of waste disposal.
5. Report any significant problems or violations of policies, practices, or procedures to the BSO as soon as reasonably possible.
6. Notify the BSO immediately if:
   a. A laboratory-acquired infection is known or suspected, or
   b. A spill of any quantity involving an agent infectious to humans, plants, or animals occurs in a public area.
7. Receive training in standard microbiological techniques.
8. Ensure that all research personnel are appropriately trained in biosafety and receive appropriate medical surveillance when needed. The PI should contact the BSO for assistance with all biosafety training needs.
9. Coordinate with the BSO and develop emergency plans for handling accidental spills and personnel contamination.
10. Create and foster an environment in the laboratory that encourages open discussion of biosafety issues, problems, and violations of procedure. The PI will not discipline or take any adverse action against any person for reporting problems or violations to the IBC, BSO, Risk Management, or State or Federal agencies.
11. Comply with shipping requirements for biohazardous agents and select agents.

In submitting proposed work to the IBC, the PI shall

1. Make an initial determination of the required levels of physical and biological containment in accordance with the requirements set forth by the NIH *Guidelines* and the CDC *publication Biosafety in Microbiological and Biomedical Laboratories* [7], as applicable.

2. Select appropriate microbiological practices and laboratory techniques to be used for the research.
3. Submit any significant changes in a given project to the BSO for review and approval.
4. Certify that the protocol does not involve dual use research.

Prior to initiating research, the PI shall coordinate with the BSO to

1. Make available to all laboratory staff and involved facilities staff (such as animal care staff) the protocols that describe the potential biohazards and the precautions to be taken.
2. Instruct and train all research personnel in:
   a. Identification of the biohazard(s) present
   b. Practices and techniques required to ensure safety and reduce potential exposure
   c. Procedures for dealing with accidents, spills, and exposures.
3. Inform the laboratory staff of the reasons and provisions for any precautionary medical practices advised or requested (e.g., vaccinations or serum collection).
4. Ensure that collaborators are made aware in advance of any biohazardous agents sent to them, and comply with all applicable packaging and shipping requirements. These materials are often regulated for shipment and must only be shipped by personnel who have received proper training and are authorized by the university to ship such materials on its behalf.
5. Maintain a formal inventory of all biological material received and sent. Logs should include the approximate quantity of the materials and where it is stored in the laboratory.

During the conduct of the research the PI shall

1. Supervise the safety performance of the laboratory staff to ensure that required safety practices are employed.
2. Investigate and report in writing to the IBC any significant problems pertaining to the operation and implementation of containment practices and procedures.
3. Immediately notify the BSO of any laboratory spills, accidents, containment failure, or violations of biosafety practice which result in the release of biohazardous agents and/or the exposure of laboratory personnel (or the public) to infectious agents. The IBC may be consulted by the BSO if necessary.
4. Correct work errors and conditions that may result in the release of biohazardous agents.
5. Ensure the integrity of all containment systems used in the project.
6. Restrict access as required by the laboratory-specific biosafety practices and procedures, and by the biosafety containment level approved by the IBC.
7. Immediately notify the BSO if a Select Agent has been isolated and confirmed from environmental and/or diagnostic specimens.

IBC member training

All new members are required to complete training on the regulatory responsibilities and functions of the IBC. This training must be completed before participation in voting activities of the committee. The IBC Chair or his/her designee will administer training. All IBC members must also complete annual retraining, covering topics that will enhance the committee's understanding of biosafety-related issues and institutional research review policies.

Regulatory gaps

Although the role of the IBC is increasingly broad, gaps in regulatory oversight still exist. The following sections will discuss these gaps and the potentials risks they introduce.

Regulatory gaps in research involving recombinant or synthetic nucleic acid research

The NIH *Guidelines* apply to all researchers at an institution if anyone at that institution accepts NIH funding for recombinant or synthetic nucleic acid research. This often means that even researchers who accept no NIH money must adhere to the NIH *Guidelines*. The NIH *Guidelines* do not apply to institutions in the United States that receive no NIH funding, however, including pharmaceutical companies, other private companies performing research with recombinant or potentially infectious materials, or private individuals. This regulatory gap allows for the possibility of risky work with recombinant DNA proceeding without any oversight.

Pharmaceutical companies, by nature, are investigating agents that will be used as therapeutics and are designed to have a favorable risk–benefit ratio. This may lead people working with these agents to believe that the agents have no hazards associated with them. However, many genetically modified study drugs are targeted against conditions with poor prognoses, so these potential therapeutics can have significant adverse side effects and still have a favorable risk–benefit ratio for the trial participant. If the clinical trial is supported solely by private funding and is performed only at clinical care centers that receive no NIH funding (e.g., private cancer centers) then the clinical trial does not fall under the NIH *Guidelines*. This means that the trial does not need to undergo public discussion through submission to NIH's Recombinant DNA Advisory Committee (RAC) or have local review by an IBC [8]. These trials would still be required to undergo Food and Drug Administration (FDA) review and local IRB review, but those bodies review different aspects than the RAC and IBC do. For example, the FDA and local IRB focus primarily on the safety of the recipients by ensuring the ethical conduct of the trial [9]. While the RAC and IBC also consider participant safety, they bring specific technical expertise needed for the assessment of the risks to close contacts of the recipient as well as the healthcare professionals preparing and administering the agent. Genetically modified viruses or other gene therapy agents may have the potential to modify the genetic code of the recipient's reproductive cells, making effective barrier contraception critical. Unlike traditional drugs which are metabolized and eventually broken down in the recipient's body, investigational agents created from microbes or modified human cells may be capable of replicating and expanding. This increases the opportunity for contamination of hospital rooms, personal bathrooms, or shared items such as eating utensils. Live agents, genetically modified or not, can pose risk of serious harm to individuals with immune deficiencies caused by organ transplant, infectious disease, pregnancy, or other conditions. Recipients of these agents must be instructed on how, and for how long, to minimize their contact with these high-risk individuals and how to properly clean and disinfect their homes to minimize the risk of transmission of potentially infectious study drug through bodily fluids. IBCs are responsible for reviewing the informed consent documents to verify that these instructions are adequately explained.

Pharmaceutical companies also produce much greater quantities of product than academic research labs, introducing different risks. Gene therapy vectors based on modified viruses (e.g., adeno-associated virus, adenovirus, or retrovirus) generally do not pose an aerosol exposure risk when used in laboratory-scale quantities. At the volumes and concentrations used in manufacturing, however, these viruses may overcome their normal route of infection and present an aerosol infection risk. In addition to the direct effects of the gene therapy product or virus itself, an accident that results in exposure to manufacturing-scale quantities may induce a strong enough immune response or allergic reaction to lead to serious illness or death. Most pharmaceutical companies choose to have safety committees, but the membership and responsibilities of these safety committees are not clearly defined and may differ greatly from company to company.

The NIH *Guidelines* also do not apply to the growing number of individuals who perform recombinant or synthetic nucleic acid experiments as a hobby. Some of these hobbyists belong to community organizations, such as DIYBio [10] or BioBricks, [11], that provide connections between amateur researchers and biosafety professionals. Others may be unaware that such safety resources are available to them or may actively avoid professional advice in pursuit of the "DIY Ethic." Sale of most equipment used in recombinant DNA research is not regulated or controlled and it is relatively simple to acquire the necessary supplies to create a modest but fully functional laboratory. Online auction sites have thousands of listings for PCR machines, electrophoresis equipment, pipettes, and other basic equipment and supplies. Instructions for genetically modifying *Escherichia coli* and other bacteria are readily discoverable through Internet search engines and various online communities exist to bring together "biohackers" to discuss projects and techniques. At least one company was recently founded to provide a way for research universities to sell their surplus reagents to other universities and community members. Unlike model rockets or soccer-playing robots, modified pathogens generated by DIY biologists may be capable of replicating and spreading to others who are not actively participating in the hobby. These genetically modified organisms may pose a larger community risk than other, more-traditional hobbies if the modifications introduce resistance to therapeutic drugs or allow the microbes to spread more easily or survive longer in the environment.

The NIH *Guidelines* outline several types of experiments that are deemed to be low enough risk that they are exempt from the requirements of the NIH *Guidelines*. Among these are purchase and transfer of transgenic animals, use of non-viral plasmids in cell culture lines, and expression of recombinant proteins in standard laboratory strains of bacteria and yeast. Some IBCs continue to review these experiments to confirm the investigator's assessment that the work is exempt and to provide institutional assurances that all recombinant DNA work is being performed responsibly. It may also be administratively easier to train investigators to submit all of their work to the IBC rather than ask them to be familiar enough with the NIH *Guidelines* to distinguish between exempt experiments and experiments that require review and approval prior to initiation.

Regulatory gaps in select agent research

Research involving microbes and toxins deemed to be the most likely agents of bioterrorist attacks is regulated in the United States by the Select Agent Regulations [12]. Before work with any of the more than 60 agents begins, the institution must write detailed safety and security plans and be inspected by the Division of Select Agents and Toxins (DSAT) to verify that the proposed safety and security practices are sufficient to contain the agent from theft, loss, or accidental release to the environment. The Select Agent Regulations have several exemptions or exclusions, however. Select agent toxins are only regulated when a single individual possesses more than a threshold amount (0.5–100 mg, depending on the toxin). Institutions may have more than the threshold amount of toxin on their premises but remain exempt from the Select Agent Regulations because no one individual has more than the threshold quantity of toxin in their possession.

Nucleic acids from select agents or hybrid viruses that combine genes from select agent viruses with non-select agent viruses may not be regulated depending on the specific viruses and genes involved. Determining whether or not the hybrid virus is regulated as a select agent can be difficult and is the responsibility of the institution performing the work (after consulting the select agent and toxin list as updated by the responsible regulatory agencies, USDA and HHS). Consequently, different institutions may come to different conclusions about what is, or isn't, regulated, and work with potentially dangerous hybrid viruses may be carried out in standard laboratory settings rather than in laboratories with the enhanced security features required by the Select Agent Regulations. Because of their expertise in recombinant DNA risk assessment, IBCs are often tasked with helping the institution's Responsible Official determine whether these chimeric viruses should be treated as regulated select agents or not. At institutions where no IBC is present, no oversight body would be available to educate investigators on the possible safety and regulatory implications of mixing genes from select agent viruses with non-regulated viruses and potentially dangerous chimeric viruses may be inadvertently created. If investigators do not realize that these chimeric viruses are more pathogenic than the parent strain then they may not use the proper safety precautions or laboratory facilities. This, in turn, could lead to occupational exposures or environmental releases that could place the surrounding community at risk.

Select agents in their natural environments are also not subject to the Select Agent Regulations. Several select agents (including *Francisella tularensis* and *Yersinia pestis*) are endemic in different areas of the United States. Investigators collecting field specimens in these areas may unwittingly be collecting and storing select agent specimens. These specimens are not regulated until the select agents are identified (e.g., through nucleic acid sequencing) but they may pose significant health risks if there is an occupational exposure while handling the specimens.

Regulatory gaps in dual use research of concern and gain-of-function research

In September 2014, the *Policy for Institutional Oversight of Life Sciences Dual Use Research of Concern* was released [13]. This policy describes the requirements for local review, approval, and oversight of dual use research of concern (DURC). DURC is described in detail elsewhere in this book but, in short, DURC refers to research that produces valuable scientific information but may also be used in warfare or terror attacks. This federal policy requires that each institution designate an Institutional Review Entity (IRE) to review DURC projects and assist in creation of risk mitigation plans for these projects. The companion guide [14] that accompanies the federal policy suggests that IBCs are the most logical choice for the IRE since similar expertise is needed to assess the risks associated with recombinant DNA research and modifications of pathogens that may increase their transmissibility or pathogenicity. IBCs may not have the security expertise needed for these reviews, however, and consultants or *ad hoc* members may need to be added to fulfill the responsibilities assigned to the IRE. The DURC policy only applies to institutions that receive federal funding for life sciences research and is restricted to research involving 15 specific pathogens. Institutions or individuals that receive no federal funding are under no obligation to follow the DURC policy, nor are individuals who are working with pathogens other than the 15 specifically listed in the DURC policy. Many common human pathogens (e.g., *Salmonella*, *E. coli*, or influenza) are not on the list of pathogens covered by the DURC policy. Consequently, there is no regulatory reason why researchers cannot modify the pathogens to become more virulent. While this research would most likely be done to study the mechanisms that make pathogens more deadly and to better understand how to control natural infections with these pathogens, publication of this research may provide a roadmap for individuals who would use biological agents in terrorist attacks.

In October 2014, the White House announced a funding pause and voluntary moratorium on so-called "gain-of-function" research for three pathogens (influenza, Middle East respiratory syndrome coronavirus, and severe acute respiratory syndrome coronavirus) [15]. While gain-of-function (GOF) is not specifically defined in the White House memo, the general consensus is that GOF refers to any experiments that increase the pathogenicity or mammalian transmissibility of these pathogens. The White House memo also does not describe how GOF research should be identified or reviewed at institutions performing infectious disease research. Because GOF is similar to DURC, it seems most likely that institutions will ask their IREs to review GOF research in addition to DURC. This will increase the workload for IBCs which will now be tasked to make decisions about GOF research in the absence of clear guidance or policy statements from the federal government regarding what is, or isn't, GOF research. This funding pause is not yet a formal policy and it currently only applies to individuals receiving federal funding for these three specific agents so the number of researchers impacted is expected to be low.

Since the DURC policy and GOF funding pause only apply to federally funded research and IBCs are only required at institutions receiving NIH funding, a potentially dangerous gap exists in the oversight of recombinant DNA research with pathogens. A properly functioning IBC will have the expertise needed to anticipate whether proposed research has the potential to increase the pathogenicity of an agent and recommend additional safeguards to protect the individual researchers and the surrounding community from an accidental exposure or release. The DURC policy and GOF funding pause provide more guidance on how and when IBCs need to formalize these types of reviews and recommendations. Institutions or individuals that receive no federal funding may not have access to the expertise needed to assess the risks associated with the modifications they are making to lower-risk pathogens and may end up creating potentially dangerous infectious agents.

Conclusions

As discussed in this chapter, IBCs play a critical role in assessing the risks associated with recombinant DNA or synthetic nucleic acid research. This role is ever-expanding and includes such diverse topics as facility design, collection of wild animals, production of engineered nanomaterials, and medical surveillance. Even with all that IBCs are asked to do, there are still several areas of research that receive no IBC review because the research falls outside of the existing regulations. These gaps create the possibility for potentially dangerous, if well-intentioned, research to continue without appropriate safety oversight. But, it is highly likely that any new mandates for oversight responsibility will fall to the IBC at most universities and research establishments.

References

[1] Russell B. Bertrand Russell Quotes. Quotes.net - STANDS4 LLC. <http://www.quotes.net/quote/7546>.
[2] National Institutes of Health – Office of Biotechnology Activities. *Requirements for IBCs under the NIH Guidelines*. US Department of Health & Human Services, 2014. Available from: <http://osp.od.nih.gov/office-biotechnology-activities/biosafety/institutional-biosafety-committees>.
[3] *Laboratory Biosafety Manual*. Raleigh, NC, USA: North Carolina State University, 2014. Available from: <http://www.ncsu.edu/ncsu/ehs/www99/left/bioSafe/index.pdf>.
[4] National Institutes of Health – Office of Biotechnology Activities. *NIH Guidelines for Research Involving Recombinant or Synthetic Nucleic Acid Molecules (NIH Guidelines)*. US Department of Health & Human Services, 2013. Available from: <http://osp.od.nih.gov/office-biotechnology-activities/biosafety/nih-guidelines>.
[5] *The North Carolina Plant Sciences Initiative: An Economic Feasibility Study*. Raleigh, NC, USA: NC State University, 2014. Available from: <https://go.ncsu.edu/plantsciencesinitiative>.
[6] *The North Carolina Food Processing and Manufacturing Initiative: An Economic Feasibility Study*. Raleigh, NC, USA: NC State University, 2014. Available from: <https://go.ncsu.edu/foodmanufacturinginitiative>.

[7] Wilson DE, Chosewood LC, editors. Biosafety in Microbiological and Biomedical Laboratories (BMBL) (5th ed.). US Department of Health and Human Services, 2009. HHS Publication No. (CDC) 21-1112. Available from: <http://www.cdc.gov/biosafety/publications/bmbl5/>.

[8] National Institutes of Health – Office of Biotechnology Activities. *FAQs about the NIH Review Process for Human Gene Transfer Trials.* US Department of Health & Human Services, 2013. Available from: <http://osp.od.nih.gov/faq/faqs-about-nih-review-process-human-gene-transfer-trials>.

[9] *Protection of Human Subjects.* US Department of Health & Human Services, 2009; 45 CFR Part 46. Available from: <http://www.hhs.gov/ohrp/humansubjects/guidance/45cfr46.html>.

[10] DIYBio – An Institution for the Do-It-Yourself Biologist. Available from: <http://diybio.org>.

[11] BioBricks Foundation – Biotechnology in the public interest. Available from: <http://biobricks.org>.

[12] *Possession, Use, and Transfer of Select Agents and Toxins.* US Department of Health & Human Services, 2005; 42 CFR Parts 72 and 73.

[13] *US Government Policy for Institutional Oversight of Life Sciences Dual Use Research of Concern.* US Government, 2014. Available from: <http://www.phe.gov/s3/dualuse/Documents/durc-policy.pdf>.

[14] National Institutes of Health. T*ools for the Identification, Assessment, Management, and Responsible Communication of Dual Use Research of Concern: A Companion Guide.* US Department of Health & Human Services, 2014. Available from: <http://www.phe.gov/s3/dualuse/Documents/durc-companion-guide.pdf>.

[15] *Doing Diligence to Assess the Risks and Benefits of Life Sciences Gain-of-Function Research.* Office of Science and Technology Policy, 2014. Available from: <http://www.whitehouse.gov/blog/2014/10/17/doing-diligence-assess-risks-and-benefits-life-sciences-gain-function-research>.

Dual use research

William S. Mellon

Chapter Outline

Historical issues

The advances in the sciences have benefitted humankind in multiple ways throughout history from enhanced health, to better agricultural practices, to numerous aspects of energy, communication, and travel. Unfortunately, with the advances for useful purposes comes the possibility that basic fundamental and applied research can be misused and result in harm to peoples and result in national/international security threats. As scientists, we believe that research should be conducted safely and, when necessary, securely. The incidence of either laboratory workers or the public being harmed accidentally is extremely small. Conscientious scientists and the meticulous oversight of life science research by Institutional Biosafety Committees (IBCs) have

Ensuring National Biosecurity. DOI: http://dx.doi.org/10.1016/B978-0-12-801885-9.00006-8

been responsible in large part for this success. IBCs review biosafety protocols and ensure safety while working with recombinant DNA and have been largely successful in this endeavor to protect researchers and ultimately the public.

While much of life sciences research has been important to many facets of life, there is none more important than the research on microorganisms and infectious diseases. These advances have led to landmark discoveries including development of drug treatments such as: penicillin by Florey [1] from Fleming's original discovery [2]; identification of streptomycin from soil, actinomycetes by Schatz et al. [3]; Jenner's vaccine work for the prevention of smallpox (reported in 1798); and Sabin's and Salk's development of the polio vaccine. In the past 50–60 years, a large volume of research on many pathogens has occurred with only rare instances of those organisms being released into the environment through accident, negligence or via direct intentions. For instance, the Centers for Disease Control and Prevention (CDC) reported 11 laboratory-acquired infections associated with "Biological Select Agents and Toxins" (BSAT) releases that were reported to the CDC between 2004 and 2010, in an average annual population of approximately 10,000 individuals with approved access to BSATs [4]. Mainly due to the conscientious efforts of scientists and the IBCs, this research is conducted more safely than at any other time in history because of the increased concern for safety and biosecurity. However, given the dangers of a subset of pathogenic materials, we must remain vigilant in the practice of biosafety and biosecurity to minimize the threat to human health, destruction of agricultural products (both plant and animal), and the potential for causing fear in the populace. Within this subset of highly pathogenic materials are 80 (as of April 4, 2013, this number has been reduced to approximately 66) or so of the most dangerous bacteria, fungi, viruses, and toxins, which have been labeled by the US Government as BSAT and are subjected to enhanced biosafety and biosecurity procedures. Data collected by the CDC between 2004 and 2010 regarding BSAT theft loss and release indicate that the risk of exposure to BSATs managed by US laboratories is low [4]. It should be noted that "release" refers to occupational exposure or to the release of a BSAT outside the primary barriers of the biocontainment area. Moreover, over 40% of the reports submitted to CDC for theft, loss or releases were submitted by entities exempt from registration (i.e., primarily diagnostic laboratories) [4]. While working with these agents poses some danger, this research is both necessary and important to our health and our security as a nation depends on our ability to understand, prevent and, if necessary, counteract the virulence or toxicities of these agents. Thus, effective biosafety and biosecurity programs in US laboratories will continue to be critical tools to effectively manage this risk [4].

Laboratory safety, while the direct purview and responsibility of laboratory scientists and institutionally of IBCs and Environmental Health and Safety (EHS) offices, has been the subject of regulation and law within the United States since the last century. The emergence of recombinant DNA technology and the concern about laboratory safety prompted a landmark decision in the early 1970s [5]. These concerns led to a voluntary moratorium by the science community until strategies for mitigating the risk could be thoroughly addressed. The Asilomar Conference of 1975 on recombinant DNA discussed the possible dangers of recombinant DNA and led to the development of tentative guidelines for the safe handling of this technology [6]. The conference attendees were diverse and consisted of scientists, legal minds, members of the press, and governmental officials. The guidelines proposed were subsequently promulgated by the National Institutes of Health (NIH) and determined how scientists would conduct recombinant DNA research.

A second significant legal event occurred worldwide with the adoption of the international treaty from the Biological and Toxin Weapons Convention (BTWC) in 1972 [7]. This treaty prohibits the development, production, and stockpiling of biological materials "of types or in quantities that have no justification for prophylactic, protective, or other peaceful purposes." Moreover, the United States internally enacted the Biological Weapons and Anti-Terrorism Act of 1989 and related statutes, policies, and guidance, which prohibit the development of biological weapons [8]. At the same time, the United States supports international research on pathogens and toxins, designated for legitimate uses and with outcomes beneficial for human, animal, and agriculture well being. Indeed, the BTWC calls upon States Parties to engage in the "fullest possible exchange" of technology and knowledge for peaceful purposes.

The events of September 11, 2001 and the intentional anthrax release incident in October 2011 have forever changed the course of how all aspects of life science research involving various pathogens are conducted. The conundrum for the US Government is how to promote basic fundamental research on dangerous pathogens and toxins that affect human, animal, and plant health, while protecting the public from the misuse of this information/material by those who would use it to do harm. Prior to 2001, the Antiterrorism and Effective Death Penalty Act of 1996 required the Secretary of Health and Human Services to promulgate regulations identifying biological agents that have the potential to pose a severe threat to public health and safety and to regulate their transfer [9]. The United States of America (USA) PATRIOT Act of 2001 established for the first-time a process for identifying and restricting individuals who could possess Select Agents [10]. While the Select Agent list was established in 1997, the provisions of the USA PATRIOT Act and subsequent modification by the *Public Health Security and Bioterrorism Preparedness and Response Act of 2002* [11] greatly expanded the regulations governing the possession of Select Agents. Moreover, the Act detailed requirements for: implementation of FBI background checks for individuals possessing these agents; entity registration; plans for physical security; appropriate biosafety to guard against accidental release and plans to respond in the event of an unforeseen accident (such as a catastrophic weather disaster), theft, or the occurrence of an accidental release. Inspections by the CDC and the Animal and Plant Health Inspection Service (APHIS) are used to assess the adequacy of the plans. The Select Agent Program began functioning in 2003 with interim rules, which were finalized in April 2005 as three sections governing human, plant, and animal agents and toxins [12].

While these new policies and regulations on possession and use of Select Agents were being refined, the concepts of biosafety versus biosecurity began to emerge in the terminology of these policies/regulations. The Select Agent Rules clearly stipulate expectations for biosafety and restricting access preventing loss, theft, or misuse. In addition, the fifth edition of the Department of Health and Human Services (DHHS) manual *Biosafety in Microbiological and Biomedical Laboratories (BMBL)* [13] aside from publishing its biosafety standards, for the first time sets standards for biosecurity; the former protecting individuals from the harmful pathogen or toxin, the latter protecting the pathogen or toxin from being possessed by non-authorized individuals [13].

Emergence of dual use research concepts/issues

In the wake of the new rules and regulations governing the use and possession of pathogens and toxins, the US Government remained engaged with the research community and continued requesting proposals through a variety of federal agencies for the development of technologies and for treatments of conditions caused by these pathogens or toxins, as well as proposals with a biodefense focus intended to protect the population and enhance national security of the United States (e.g., Biodefense and Emerging Infectious Disease Research Opportunities 2005/2006 [NIAID]). There was, however, another emerging concern: that new information generated from this research could be misapplied and pose significant risk to national security. Thus, the National Research Council (NRC) was tasked by the US National Academies to examine the paths by which advances in pathogen and toxin research could be utilized for nefarious purposes.

In 2004, the NRC published a report entitled Biotechnology Research in the Age of Bioterrorism [14]. The report recognized that the advances in molecular biology and genetics have revolutionized many aspects of life and agricultural sciences to the general benefit of society. However, these advances also pose a significant risk because they could be used to create the next generation of biological weapons. Moreover, given the availability of rapid and uncontrolled communication via the Internet, the likelihood of misuse of scientific information is exacerbated. The Committee made a number of recommendations that would establish stages at which proposed or completed research would be reviewed to ensure consideration was given to the possibility of applications for bioterrorism. The system proposed by the NRC report relied heavily on self-governance by the research community and on the expansion of the existing regulatory process. Within the framework of these recommendations, the Committee listed seven categories of experiments involving microbial agents with particular potential for misuse, described below. They recommended that the biosafety application for experiments that fall within the seven areas of concern be especially addressed by the researcher and reviewed by the IBC of the institution so as to promote careful evaluation of potential benefits versus potential danger. Among these recommendations is also the formation of a federal advisory group that could act as a resource by providing guidance and leadership for the system of review and oversight where dual use research is identified. Subsequently, in 2004 and upon the recommendation of the NRC, the US Government created the National Science Advisory Board for Biosecurity (NSABB).

Review of the process for rulemaking and oversight

The role of the NSABB

One of the fundamental tasks of the NSABB was to develop criteria for identifying dual use research of concern (DURC). The proposed criterion is "research that, based on current understanding, can be reasonably anticipated to provide knowledge, products, or technologies that could be directly misapplied by others to pose a threat to public health and safety, agricultural crops and other plants, animals, the environment, or material" [15]. The NSABB was also charged with "proposing and developing oversight framework for the identification, review, conduct, and communication of life sciences research with dual use potential" [15], taking into consideration both national security issues and the needs of the life sciences research community in their research and publication efforts to advance society's needs. The mutual conflict inherent in these efforts is the central dilemma of dual use research. NSABB has developed guiding principles on oversight of dual use research, a proposed system that includes responsibilities for the federal government, institutions, and researchers. Most importantly, local oversight is stressed and suggested steps for implementation of dual use research review are demonstrated in Figure 6.1.

NSABB developed seven categories of conditions, using the NRC report as a starting point, which meets the criterion developed to define DURC. Each of these categories is provided (see below) and NSABB provides clarifying comments or examples [15]:

1. Enhance the harmful consequences of a biological agent or toxin.
2. Disrupt immunity or the effectiveness of an immunization without clinical and/or agricultural justification.

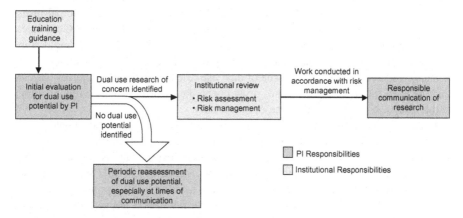

Figure 6.1 Major steps in local oversight of dual use life sciences research.
Permission granted from NIH – A Report of the National Science Advisory Board for Biosecurity, June 2007, see Ref. [15].

3. Confer to a biological agent or toxin, resistance to clinically and/or agriculturally useful prophylactic or therapeutic intervention against that agent or toxin or facilitate their ability to evade detection methodologies.
4. Increase the stability, transmissibility, or the ability to disseminate a biological agent or toxin.
5. Alter the host range or tropism of a biological agent or toxin.
6. Enhance the susceptibility of a host population.
7. Generate a novel pathogenic agent or toxin or reconstitute an eradicated or extinct biological agent.

US Government response

The regulatory environment has evolved over recent decades, more so with the advances in life sciences technologies and with the realization that terrorists could misappropriate new research results meant for the benefit of humankind. The United States Department of Agriculture (USDA), NIH, CDC, Food and Drug Administration (FDA), Department of Defense (DOD), Department of Energy (DOE), and National Science Foundation (NSF) all have links to life sciences research in one way or another and have regulatory roles with respect to research that is carried out and published. The NRC report of 2004 [14] (see Chapter 2) aptly summarizes the regulatory climate prior to September 11, 2001. Most of the efforts by the organizations listed above were focused on protecting the public health, safeguarding workers and the general environment from accidental release, and ensuring containment of recombinant molecules. Notable examples of government intervention were that the NIH published safety guidelines in 1976 [16] after the Asilomar Conference followed by the publication in 1984 of *BMBL* [17]. The *BMBL* guidelines address laboratory safety procedures for working with and handling infectious disease agents and categorize infectious agents and laboratory activities into four levels (BSL-1 to BSL-4) (see Chapter 2). The NIH Office of Biotechnology Activities, the Recombinant DNA Advisory Committee (RAC), IBCs at individual institutions, and the principal investigators (PIs) themselves share responsibility for enforcement of the NIH Guidelines. Post September 2001, the US Government has enacted a variety of rules/regulations and oversight structure (depicted above) that govern the control of information in life sciences research. The evolving issue of dual use research in the life sciences has created the need for greater oversight of researchers, their research, and communications of the results, especially in the area of biosecurity.

Following more than a decade of policy, security, and scientific debate in the midst of the H5N1 avian influenza publication controversy from the laboratories in Madison, WI and Rotterdam, Netherlands, the US Government issued an overarching federal policy on the oversight of DURC on March 29, 2012 [18], although not as a direct result of the avian flu controversy. The March 2012 policy codifies the definition of DURC, is applicable to extramural or intramural research that is funded by a federal department or agency, identifies 15 agents or toxins derived from the Select Agent list, and coincides with the seven categories of experiments (although modified slightly) defined by NSABB. It further directs federal departments and agencies, which conduct or fund life sciences research, to review their research portfolios to determine whether research projects fall within the scope of this policy and to assess the risk and benefits of such projects, and also to develop risk mitigation and communication plans in collaboration with the institution and researcher. Should research rise to the level of DURC, the mitigation strategy to reduce risk can include modifying the research, enhancing biosafety and biosecurity, and/or developing communication strategies to address concerns about the potential misuse of the proposed research, information, technologies, and knowledge.

The US Government released a second draft component of the DURC policy, aimed at strengthening ongoing institutional review, oversight, and investigators, in the Federal Register on February 22, 2013 with an open comment period of 60 days [19]. The comment period was restricted to a set of 16 questions posed to the research community concerning the implementation of DURC at institutions. Prior to the release of the proposed 2013 DURC Policy, the US DHHS also released additional guidance on a particular subset of DURC [20]. The DHHS Framework entitled "A Framework for Guiding U.S. Department of Health and Human Services Funding Decisions about Research Proposals with the Potential for Generating Highly Pathogenic Avian Influenza H5N1 Viruses that are Transmissible among Mammals by Respiratory Droplets" resulted in an additional layer of oversight imposed on proposals for research anticipated to generate highly pathogenic avian influenza H5N1 viruses that are transmissible in mammals by respiratory droplets. The new Framework outlines a rigorous review process that takes into account the scientific and public health benefits of the research, the biosafety and biosecurity risks, and the appropriate risk mitigation measures appropriate for the proposed research. The framework for guiding DHHS on funding decisions applies to proposals that are reasonably anticipated to confer gain-of-function attributes that enable influenza viruses expressing the virulent form of the hemagglutinin gene from highly pathogenic H5N1 to be transmissible among mammals by respiratory droplets. A proposal can only be forwarded for consideration if it meets the seven criteria proposed in the Framework and has been successfully considered by the funding agency's Scientific Merit Review and Dual Use Review. Once a proposal meets these criteria it is subjected to further review by DHHS to determine whether the proposal is acceptable for DHHS funding. For NIH awards, DHHS review will occur before NIH council review.

The proposed 2013 DURC Policy has now been finalized as of September 24, 2014 with an effective start date of September 24, 2015 [21] (referred to in this chapter as the 2015 DURC Policy) at which time all entities must establish procedures necessary to comply with this policy. The 2015 DURC Policy applies to all federal agencies, as well as to institutions that receive federal funding or that are conducting research that meets the definition of DURC, regardless of funding source. Like the March 2012 policy, the new policy limits the scope of DURC to 15 Select Agents/Toxins and seven categories of experiments. Institutions to which this policy applies are required to certify at the time of seeking funding after September 24, 2015, that they are in compliance with all aspects of this policy. Non-compliance with this policy could result in loss of federal funding for the institution.

The researcher has the primary responsibility for identifying potential DURC based on using one of the 15 select agents. Upon performing research that produces, aims to produce, or can be reasonably anticipated to produce one or more of the seven categories of experiments; or based on other criteria that would make the research fit the scope of the DURC policy, they must initiate the review process. However, it is not clear whether the review process by the researcher and the institution should be initiated during early stages of proposal development and/or prior to submission of the research plan to a funding agency. Given the difficulties in obtaining federal research funding at the present time, it would seem wasteful to carry out a significant assessment until peer review has occurred and funding is likely. Moreover, there are requirements for DURC education of all research personnel, including students, which further creates an unnecessary burden if funding is not secured. Once research is ongoing and results are to be communicated, the DURC 2015 Policy indicates that the communication must be made in a responsible manner without any guidance as to what constitutes "responsible." Institutions will need to address these issues prior to September 24, 2015 in a way that satisfies the spirit of the DURC 2015 Policy and is consistent with the "academic culture" of the institution.

Institutions will be required under the DURC 2015 Policy to construct a review process for assessing whether research that uses one or more of the 15 agents or toxins is, in fact, DURC. While the 2013 proposed policy did not stipulate how the review was to be accomplished, the DURC 2015 Policy requires a committee established by the institution, referred to as the Institutional Review Entity (IRE), which is empowered to execute the requirements of this policy. The identification by the PI of research requiring DURC review is the trigger for initial referral to the review committee, which is depicted in Figure 6.2 articulating the process in the DURC 2015 Policy.

Institutions are required to designate an Institutional Contact for Dual Use Research (ICDUR), to serve as an internal resource for issues regarding compliance and the implementation of the requirements for the oversight of research that meets the definition of DURC. The ICDUR also serves as the liaison between the institution and the relevant program officers at the federal funding agency, or for non-federally funded research, between the institution and NIH (or the appropriate federal funding agency to which NIH refers the institution). Within 30 calendar days of identifying research falling within the scope for DURC potential, whether it ultimately meets the definition of DURC or not, institutions need to notify the federal funding agency and/ or NIH. Additionally, within 90 calendar days from the time the institution determines a line of research to be DURC, the institution must provide a copy of the risk mitigation plan to the funding agency for review. Institutions are also required to make sure that those working with DURC have received appropriate education and training about DURC and risk mitigation. They also need to make sure that there is a mechanism for reporting instances of non-compliance and a mechanism for researchers to appeal institutional findings of DURC. Finally, institutions must provide an annual assurance of compliance to the government.

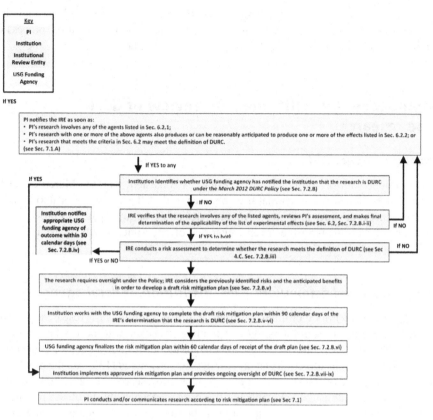

Figure 6.2 An overview of the process for institutional review of life sciences research in the DURC 2015 Policy.
Permission granted from NIH.

Other institutions (NIH, academic)

The response by NIH to the US Government's 2012 DURC Policy was immediate in that agencies reviewed all funded proposals in their portfolio to determine which proposals might have DURC aspects. Institutions were contacted to initiate the process delineated in the DURC policy. Reviews by institutions were carried out within the timelines stated within the DURC policies. Once the federal funding agency approved the DURC and mitigation plan, the "terms and conditions" of an existing grant were changed to meet the new DURC policy. While there is no precise measure published of the number of entities that were contacted nor to their readiness to respond, the observations of this author from attending several AAAS/AAU/APLU/FBI sponsored meetings (September 13–14, 2012) [22] would suggest that there was a range of readiness and experience. Some institutions, such as the University of Wisconsin-Madison and

Boston University, had been carrying out DURC analysis and processes for a number of years and had a formal program in place. Other institutions were in the process of doing so, while others had only discussed and began to plan for these outcomes.

Approaches by institutions for review of dual use research

Scope of oversight

As mentioned above, the DURC 2015 Policy is consistent with the March 29, 2012 Policy, life sciences research that uses one or more of the 15 agents or toxins listed and produces, or aims to produce, or can reasonably be anticipated to produce one or more of the effects in the seven categories of experiments will be evaluated for DURC potential.

Organizational framework – PIs, institution, federal agency

The DURC 2015 Policy outlines the roles and responsibilities for researchers (principle investigators), institutions, federal funding agencies, and the US Government. It has long been recognized that researchers are the most knowledgeable concerning their research and its direction and, as NSABB so aptly noted [15], "are the most critical element in the oversight of dual use life sciences research." In fact, most reviews that have been published acknowledge the primary responsibility of the researcher for reviewing, crafting, and anticipating whether their research generates knowledge, products, or technologies that might have the potential for misuse. For this reason, the researcher and institutions invest in educating all researchers in the life sciences as to the fundamental principles of dual use research so that those conducting research can be cognizant of this concept. The proposed February 2013 Policy limited significantly the role of the researcher in the assessment process to merely identifying whether the research involved one or more of the 15 agents. The revision to the final DURC 2015 Policy retains the researcher throughout the assessment process that is more consistent with the discussion of many thoughtful groups and panels who have considered DURC regulation.

To be clear, the researcher whose research uses one or more of the 15 agents must assess whether the research also produces, aims to produce, or is reasonably anticipated to produce one or more of the seven categories of effects or may otherwise meet the definition of DURC. The researcher works with the IRE to assess the dual use risks and benefits of DURC and develop risk mitigation measures. Moreover, the researcher must conduct DURC in accordance with the risk mitigation plan, be knowledgeable of US Government policies, and ensure laboratory personnel are trained and communicate DURC in a responsible manner. The institution must notify the federal funding agency for non-federally funded research, NIH, of the review process and provision of the risk mitigation plan.

Burden to IBCs

As already indicated, the DURC 2015 Policy requires the institution to create an IRE to execute the requirements of the Policy. There is a range of options as long as the review entity is appropriately constituted and authorized by the institution to conduct the dual use review. As an aside, no mention of additional authority for the review entity is made such as securing the necessary financial resources to administer this process, ensuring cooperation from all aspects of the institution, etc. This is a necessary component, which will be addressed below. The options include: establishing a committee solely for dual use review; an extant committee such as an IBC that may or may not need *ad hoc* members depending on its current composition and whether it has the expertise to review dual use proposals; or an externally administered committee (another institution's committee or a commercial entity).

The institution will likely establish a review entity based on a number of factors including size of the research faculty and staff, level of expertise among the faculty and staff, the financial wherewithal to operate another administrative oversight committee, and the culture of the institution with regards to compliance activities. Given the rapidity with which these two US Government DURC policies were brought forth, a majority of research institutions have opted to employ the IBC and broaden their charge to now include review and oversight of DURC. The IBC will have a collection of researchers that are, in general, prepared to judge the biosafety issues associated with the Select Agents that fall beneath the DURC policies. Their ongoing processes can be extended into the dual use arena and, most probably, have an infrastructure that can be adapted to this role. Certain DURC policy requirements will be an expansion of their current tasks such as:

1. Providing education and training to researchers in the life sciences.
2. Maintaining records of personnel training.
3. Reporting instances of non-compliance and remedying the gaps in execution of the policies.
4. Engaging in dialog with researchers of the research questions and assessing risk and benefit aspects.
5. Assisting researchers when questions arise about whether their research may require further review and oversight.
6. Development of mitigation strategies for conducting DURC.
7. Maintaining records of institutional DURC reviews and completed risk mitigation plans.
8. Providing formal assurances to the federal funding agencies that the institution is compliant with all aspects of the DURC policy.
9. Establishing an appeal process internally for researchers to appeal institutional decisions.
10. Making its procedures for reviewing life sciences research for dual use potential accessible to the public.

However, there are several aspects of the DURC regulations that IBCs may not be equipped to manage without adding additional personnel. Aside from having the appropriate subset of specific research expertise to deal with DURC research issues, the members may not have either the depth or breadth of expertise to handle biosecurity issues. Since this is in part a significant focus of DURC, many IBCs may have to recruit individuals who can provide this knowledge. Moreover, biosecurity is a costly venture for institutions and most IBCs may not be routinely involved with high-level institutional administrators with regard to budgetary issues that require institution-wide decisions. The biosecurity issues for BSAT are wide-ranging and involve back-ground checks for personnel, physical security, as well as secured record-keeping and electronic communication. Thus, information technology becomes an integral part of the DURC process since the agents and toxin are also regulated under the BSAT Regulations.

The institutional reviews and mitigation plans must be communicated to the appropriate federal funding agencies in a timely manner. The DURC 2015 Policy requires the establishment of a designated official, the ICDUR. This individual serves as an internal resource to the institution with regards to compliance issues and implementation of the oversight process. In addition, the ICDUR acts as a liaison to the federal funding agencies in communicating mitigation plans and resolving issues between the researcher, institution, and federal funding agencies. Does the chair of the IBC play this role, and if not who does? Lastly, the communication of DURC to the scientific community is challenging. The March 29, 2012 Policy gives the federal agency the purview to determine the venue and mode of communicating (address-ing content, timing, and possibly the extent of distribution of the information) the research responsibly. This also involves mitigating the risks adequately. This becomes a tripartite process with the main roles being carried out by the researcher and federal funding agency, but also involving the institutional entity as it is related to the agreed-upon mitigation plan. Thus, this requires exceptional skills by the IBC members and the ICDUR.

Other models

Institutions will consider the best model that will permit compliance with the DURC Policies. One of several mechanisms may accomplish the required tasks within the policies. An example of how the University of Wisconsin-Madison approached this issue and why is now presented; it is not necessarily better than other approaches, but it works extremely well for this institution.

During the aftermath of September 11, 2001, the University of Wisconsin-Madison established an Access Control Committee (ACC) to triage the campus buildings/ facilities according to the need for enhanced security. Moreover, as the Select Agent Rules were being formulated, an offshoot of the ACC was formed to oversee the Select Agent Program at the University of Wisconsin-Madison. This Committee, the Biosecurity Task Force (BTF), was charged with preparing the campus for imple-menting processes to meet the requirements of the finalized regulations. Once the Select Agent Program was established, the BTF was charged further with overall responsibility for oversight of the Select Agent Program. The BTF was formed to deal with Select Agent Program oversight rather than expanding the IBC for several reasons. UW-Madison officials determined that biosecurity would require a very significant and continual commitment of time and resources, the Select Agent

Program on campus would be large based on the number of projected PIs, and a committee devoted solely to oversight of the Select Agent Program would demonstrate clearly its commitment to the newly mandated rules. Mainly for these reasons the IBC was not tasked with BSAT review as it already had a high workload and the BTF was finalized. Thus, the oversight by the BTF includes approval of all principle investigators, the high-containment facilities, biosecurity, and to be advisory to the Responsible Official. The current membership of the Committee was expanded in 2010 while I was the Responsible Official as well as serving as the Associate Dean for Research Policy in the Graduate School. Below is the composition of the Committee as of December 2013, which includes representatives from various sectors of the campus:

- Select Agents Responsible Official (Chair)
- The two Alternate Responsible Officials (AROs)
- Director of Environment Health and Safety
- Biosafety Officer
- The Associate Deans for Research from the three schools/colleges representing academic homes of the select agent PIs
- State Lab of Hygiene member who is also a Select Agent PI
- The Director of the Veterinary Diagnostic Laboratory
- Two officers from the UW Police Department
- Campus Attorney whose expertise is in research policy
- Director of Campus Information Security
- Director of University Health Services Physician who adds occupational health expertise
- Two members from the UW Communications Department.

The formation of this Committee had immediate benefits to the campus and its Select Agent Program. It complements the IBC by providing expertise not represented on the IBC as both AROs are *ad hoc* members of the IBC and the Biosafety Officer is a member of the BTF. This overlap in memberships permits complementary exchange of information during the discussion of select agent biosafety protocols during the IBC meetings. The BTF added assistance in the planning and construction of the Influenza Research Institute to meet the needs of future highly pathogenic research efforts. To date, the cost of construction and improvements to the Influenza Research Institute totals nearly $14 million but permits research to be conducted at Biosafety Level 3 enhanced and Agriculture Biosafety Level 3, which enhances overall biosafety and biosecurity for a variety of research. The BTF also implemented discussions and review for DURC early in the planning of transmission experiments involving avian influenza (H5N1) and prior to submission of the research by the Kawaoka team that was subsequently published in *Nature* [23]. It came as somewhat of a surprise to many of us that had been involved in both the research and oversight when NSABB initially recommended omitting key details from the H5N1 transmission studies prior to publication. Our surprise was based on our prior knowledge of the elevated biosafety and biosecurity conditions that these experiments were carried out under, the thoughtful and thorough discussions concerning dual use issues that had occurred, and finally the importance of these results to the influenza research and surveillance communities. Complicating matters though was the fact that NSABB was not only considering the Kawaoka/*Nature* manuscript but also concomitantly the manuscript from the Dutch group in Rotterdam whose results and methodological approaches provided other novel outcomes [24]. For several months in the spring of 2012, dual use research issues involving H5N1 were the topic of many articles, commentaries, and discussion, which eventually led to an international investigator self-imposed moratorium of influenza transmission study research. Moreover, complications occurred when the World Health Organization (WHO) invited selected participants to Geneva (February 16–17, 2012) to discuss [25] differing opinions surrounding this research. Key influenza researchers, editors of both journals, heads of WHO surveillance centers, countries who provide influenza viruses, NIH, and chair of NSABB, attended the meeting. However, a secondary issue arose as the Department of Commerce concluded that since *Nature*/NIH had voluntarily agreed to withhold certain details from the *Nature* publication and that the WHO meeting was not open to the public, the publication restriction invalidated the fundamental research exception with respect to portions not already in the public domain (EAR § 734, discussed in Supplement 1). Subsequently, NIH obtained an export license so Americans could discuss at WHO and Japanese export law required Dr Kawaoka to obtain a license to travel from Japan, where he was at the time of the decision, to Geneva. The WHO meeting had several important outcomes but none more important than "the results of this new research have made it clear that H5N1 viruses have the potential to transmit more easily between humans underscoring the critical importance of continued surveillance and research with this virus."

Based on the NSABB's recommendations on oversight of "dual use research" and catalyzed by the recent events associated with the influenza papers, the US Government released its guidelines for the life sciences community to help them identify and minimize the risk that biological research or knowledge could be misused to cause harm [18]. Following the March 29, 2012 DURC Policy issuance, the University of Wisconsin-Madison developed a more formal process for DURC review and oversight to meet the requirements of this policy. Identification of "DURC" involving Select Agent and Toxins research by PIs, internal university committees, and/or an external funding agency initiates a formal review and oversight process at the University of Wisconsin-Madison.

1. The PI and his/her selected laboratory personnel meet with the University's Select Agent and Toxins Responsible Officials to discuss the relevant issues of "DURC" based on the seven categories of experiments established by NSABB. The information gathered during this process is used as the basis for the assessment, and includes, the research protocols, grant aims, progress reports, and any other information needed to assist in the University's review of the research for its dual use potential.

2. These documents are forwarded to the chair of the IBC, who designates a subcommittee consisting of approximately three IBC members to interview the PI and assess the potential "DURC." Based on their evaluation of the documents and discussions with the PI, the IBC subcommittee writes a report that reviews the grant and its aims, describes the documents provided by the PI for review, assesses whether the research grant aims contain potential "DURC" based on the NSABB's seven categories of experiments, and makes a formal recommendation about whether and how to proceed.

3. The IBC subcommittee report is presented to the full IBC for discussion and vote.

4. The full IBC votes and the report are forwarded to the BTF, for further discussion and vote. In some instances, the BTF sends recommendations or questions back to the IBC for consideration. The process continues until the members of both the IBC and Biosafety Task Force agree on the conclusions of the review and provides the PI their final response. This process involves input from the PI.

5. The Select Agent Responsible Official assumed the lead role for the coordination of the DURC review and provides the federal funding agency a letter explaining the review process, the final decision of whether the research is "DURC," and the IBC subcommittee report [26].

Following this, the WHO recommended that publication of the papers should be delayed rather than be published immediately in partial form [25]. After the authors revised their papers to address the security concerns, the NSABB reviewed the papers again and concluded that the information contained in the revised versions no longer posed security risks and supported publication of the revised papers [27]. Subsequently the dual use issues were resolved between Dr Kawaoka, the editors of *Nature* and NIH and the H5N1 transmission studies were published in June 2012 with enhanced and expanded biosafety and biosecurity methods described [23]. The process outlined above has worked well for the University of Wisconsin-Madison as four NIH-funded grant proposals needed immediate review under the March 29, 2012 DURC Policy within the initial 6 months of these regulations.

Future of biosecurity implementation

A series of meetings sponsored by the American Association for the Advancement of Science (AAAS), the Association of American Universities (AAU), Association of Public Land-Grant Universities (APLU), and the Federal Bureau of Investigation (FBI) Weapons of Mass Destruction Directorate (WMDD) have been held over the last few years to strengthen relationships between the various stakeholders for improving both local and national-level biosecurity issues. The meetings were held as not-for-attribution to encourage open dialog and the reports have been prepared by AAAS and published to summarize the major themes and highlights [22]. Overall, participants highlighted the importance of a strong supportive institutional environment in order to implement the high standards of biosecurity needed to satisfy the DURC policies vis-a-vis the Select Agent Regulations; in the absence of such an environment, administrative staff, researchers, and public health officials face serious challenges in implementing these regulations. Whether the institution is primarily academic or industrial, the financial cost involved in establishing and maintaining a robust staff, administrative support, and safe and secure facilities is high. The financial burden mounts as academic institutions simultaneously face decreased funding from their state, increased regulatory requirements for a wide range of issues, and very little to no help from grants that support select agent/DURC research for personnel and physical security upgrades.

Proposed regulation burdens

While academic institutions have faced increased regulatory and financial burdens with the installation of the latest DURC policies, they have also experienced increasing compliance requirements (beyond security requirements) on biological research, while having to contend with an indirect cost rate that has remained at the same level since the early 1990s [28]. These requirements encompass a wide range of issues associated with garnering of federal research monies including, managing conflicts of interest, research integrity, export control, human and animal subjects protection, to delineate but a few. Many universities spend significant administrative and financial resources to comply with guidelines, regulations, and laws that apply to biological research. During the past several years, the AAU and APLU have initiated efforts to better understand the total financial costs of conducting research at educational institutions [29]. AAU and APLU have found that the cost of complying with unfunded research requirements has, in several cases, doubled or even quadrupled over the past 5 years. In addition, several universities cited a significant amount of their general operating funds going towards research compliance. This redirection of support away from education, general facility maintenance, direct costs of research, and other core university functions may adversely affect the entire education and research environment at academic institutions.

Antithesis of science for good/misuse

At the heart of the issue with DURC is the fact that while the vast majority of biological research is intended for the betterment of humans and animals, some can be used by a few for nefarious purposes. This has led to the current precautionary regulatory environment and new awareness by responsible scientists regarding the potential consequences of their research if it were misapplied. The ongoing controversy with influenza research has focused on the gain-of-function experiments and the potential risk associated with this research. The University of Wisconsin-Madison approved conducting such experiments as the risks could be mitigated by implementing rigorous measures of biosafety and biosecurity while permitting enhanced understanding of the transmission mechanisms of the virus and the potential development of treatment modalities against the virus. Moreover, given the larger picture that most microbiological research has the potential to be carried out using gain-of-function experiments to advance the science of microbiology and treatment of infectious disease, these experiments seem relevant to the consideration of DURC on a larger scale. While the instances of noted misuses have not occurred at academic institutions or as a result of the research at these institutions, the regulatory response seems to focus disproportionately on these institutions because they receive the majority of the federal funding supporting biological research with DURC potential. The long-range question will be whether well-intentioned scientists and the institutions will continue to invest time, energy, and resources into research that has DURC potential if the additional regulatory burden is not fully funded?

Communication, protection versus dissemination

Aside from the actual products of DURC, is the dissemination of the information that is produced during its discovery, which may be misused. Thus the DURC 2015 Policy will require approval by the federal funding agency or NIH in cases of non-federal-funded research of the risk mitigation plans, including all communications that fall within the limits of the risk mitigation plan. The University of Wisconsin-Madison's experiences to date under the existing DURC 2012 Policy is one of tri-partite collaboration including NIH, the investigator, and the institution. Explicating biosafety and biosecurity safeguards is paramount in order to mitigate risks to the environment and lives. The experience with H5N1 publications involving gain-of-function experiments would suggest that there is a rational middle ground in which the benefits of this research when carried out under rigorous conditions have been judged to outweigh the potential risks. The goal that we sought was and continues to be to allow conduct of crucial research where the asymptote of risk is nearly zero or as close to zero as possible.

Necessary partnerships (US Government, funding agencies, PIs, institutions)

It is important for the various stakeholders (scientists, policymakers, security experts, and industry and academic leaders) to have meaningful discussions concerning all aspects of dual use research, so that whatever rules/regulations are imposed by the federal government, do not under- or overestimate the security needs to ensure national security. Soliciting a wide range of viewpoints can, at the same time, augment advances in basic biological sciences while enhancing national security. To this end the meetings sponsored by AAAS, AAU, APLU, and the FBI have been success-ful in deepening the understanding of the issues related to carrying out life sciences research and the security needs of the US Government. Research and security efforts cannot be conducted in a vacuum given the global nature of disease transmission and terrorism as it relates to DURC issues. If we are to be successful in improving the health of worldwide populations and fighting terrorist actions, it is imperative to continue to expand these efforts on an international scale.

In the current economic and security climate, we are reminded that universities in the United States are challenged to provide a safe and secure environment for scientists to conduct high-quality research. Research universities operate in partner-ship with the federal government and private industry, accepting funds to conduct basic and applied research, to the ultimate benefit of society. The average research university is a complex institution carrying out hundreds of research projects that involve scientists from diverse disciplines (e.g., physical, life, computer, engineering, mathematical, chemical, and social sciences). Moreover, research may involve more than one institution and may include international entities as well. The scientific staff may include multinational individuals who may or may not be in training as undergraduates, graduate students, and postdoctoral researchers. These partnerships with the government and industry have become more costly and complex over time; over the past 30–40 years, concerns about ethics, safety, security, and intellectual property related to research have led to the creation of laws, regulations, and guidance to minimize or eliminate those concerns. Since the early 1990s, the number of compliance requirements for research programs has steadily increased, with a sizable jump after the 2001 terrorist attacks, but the amount of indirect costs allotted to universities to comply with those requirements has remained constant [28,29]. We must find a mechanism to advance science to deal with a variety of maladies and protect the world's populations from untoward efforts by a few with the intent of creating harm from new scientific knowledge.

International cooperation

The US Government DURC policies will address potential issues of national security related to funded research within the borders of the United States but as science efforts are global, the question remains as to whether there are adequate controls internationally to prevent DURC products or information from being misused? Without a coordinated effort by international bodies, this seems unlikely. The growing concern over the potential use of biological agents as weapons has prompted interest in establishing national biological laboratory biosafety and biosecurity monitoring programs as discussed in the article by Blaine [30]. The US Government does not believe that an inspection regime will allow for verification that all countries are living up to their obligations under the Biological and Toxin Weapons Convention [31]. Inspection data obtained from CDC and APHIS inspections of US laboratories would unlikely be privy to any international monitoring regime, however, greater transparency and applying the compliance methodology and rules, which are "tried and true" norms should be adopted by any international community as a confidence-building measure in the Convention. Additionally, the Blaine article discussion points out the challenges and issues that should be considered by governments, or organizations, embarking on this path [30].

The meetings sponsored by AAAS, AAU, APLU, and the FBI previously mentioned have been aimed at bridging science and security for biological research. The premise of these meetings is based upon the concept that active communication between universities and the FBI could help maintain the United States' competitive advantage in research and education by helping to mitigate potential domestic and national security risks. A third meeting was held in February 2013 and focused on critical issues resulting from foreign scientists studying or working in the United States, international collaboration, and US scientists working in foreign countries [32]. While not addressing DURC issues *per se*, suggestions from these discussions could be applicable to the development of DURC policies internationally.

At present there is a limited amount of DURC research resulting from collaboration between US and non-US scientists, much of which is population surveillance intended to isolate novel influenza strains in animals. Any international work with Select Agents would require US-based researchers to include their foreign colleagues in the formal registration process, consistent with DURC-related biosafety and biosecurity measures. Research conducted abroad that would meet definitions for DURC may or may not be regulated as in the United States. Thus, closer international cooperation is required to create similar oversight to mitigate risks from illicit (or terrorist) use.

One basic issue that arises in the multinational arena is simply definitions of terms used in describing biosafety and biosecurity concepts. "The concepts of biosafety (i.e., measures to prevent accidental exposure or release of pathogens and toxins) and biosecurity (i.e., measures to prevent intentional theft, loss, or misuse of materials, technology, or expertise to cause deliberate harm) often get lost in translation from English because many languages have one word to describe both terms. To address this difference, many national and international organizations have coined the term 'biorisk management' to jointly describe biosafety and biosecurity" [32,33]. While the expression "biorisk management" clarifies the terminology and describes ways and development of strategies to minimize the likelihood of occurrence of biorisks, it does not specifically address biosecurity from a law enforcement perspective which encompasses the broad range of security threats institutions face, beyond pathogen- or toxin-specific risk assessment. These threats include personnel security and the "insider threat"; theft, loss, and/or misuse of biological pathogens or toxins; cyber attack; animal rights extremism; theft of intellectual property; and bioterrorism. Thus, the international efforts by biosafety organizations to promote biorisk management does not necessarily address the issues of DURC policy nor promote the need to prevent bioterrorism or the development of bioweapons. There is a pressing need to develop international strategies based on a common understanding of DURC issues that can be addressed in a cost-effective manner, which will lead to the development of "best practices" on an international scale.

Conclusions

The need for dual use research

The history of infectious disease outbreaks involving plague and influenza as examples, which killed millions of individuals, serves as an important reminder of the need for research to develop drugs and vaccines to prevent death and suffering. Unfortunately, a few would use the advances in biotechnology and infectious disease research to inflict harm. It is incumbent upon biological scientists, research administrators, security experts, and policy-makers to cooperate and facilitate the development of policies and programs at the local, national, and international levels to both enable scientific progress and minimize risk without inappropriate classification or inordinate restriction of biodefense and infectious disease research. If science is hobbled because of the perception of unmitigated security risks, ultimately we will pay a significant price in terms of human and animal suffering, and loss of life. This is unacceptable and international efforts are necessary to ensure that this does not occur.

The need for transparency and public trust

There is a need to strike a balance between advancing science and maintaining security, both for ensuring public safety and the need to retain public support for research endeavors with highly pathogenic and dangerous organisms. It is incumbent upon researchers to carry out beneficial research that falls under DURC but at the same time garners the public's support through behaviors that promote the highest levels of integrity, ethics, and security. The public plays an important role in both supporting research endeavors carried out by government-funded efforts and their approval of the types of research conducted. While the public benefits from this research, scientists must maintain the public trust throughout the process by acting responsibly so as to not harm the public in any way. Gaining and maintaining that trust is paramount if publically funded research is to continue and ensuring that policy-makers make research a priority for the nation. For this purpose, the University of Wisconsin-Madison was proactive in its construction of the Influenza Research Institute by inviting the public to view the facility prior to its opening and commissioning for research with highly pathogenic organisms. Our experience with this open approach has been extremely useful in garnering public support for the research efforts housed in this facility. Periodic invitations to the press, public officials, lawmakers, and selected citizens during periodic maintenance shut downs has re-enforced public trust in our research efforts. Thus, the experience of the University of Wisconsin-Madison is that active engagement and communication with the public and demonstration that scientists hear and address many of the public's concerns, including addressing biosecurity concerns, are key components of building and maintaining public trust.

Will dual use research mandates be adequately resourced?

Despite the responsibilities placed on scientists and the promise that biological research can address the health and agricultural needs of the nation while maintaining national security, the financial and administrative costs involved in ensuring that institutions and scientists are fully compliant with the rules and regulations that govern biological research with DURC potential are significant. Research universities are currently facing financial difficulties as several states have decreased their funding to higher education, the federal government has reduced their funding for research, and compliance requirements have continually increased for many areas of research. Commentaries at sponsored meetings [28,29] indicate that many public academic institutions faced with the unique combination of deceasing budgets, increased regulations, and increasing and disproportionate demands on scientists' and administrators' time with compliance activities are prompting a shift in research agendas from Select Agents and DURC-related research to other research areas that are less burdensome. If this becomes a national trend, it will have dire consequences for public health and agriculture both in the United States and possibly internationally. Moreover, it will diminish national security efforts and put us significantly behind others. It is necessary that scientists, policymakers, and the public not only support basic and applied research at an appropriate level, but that infrastructure, compliance needs, and security be financially supported at an appropriate level as well to maximize national security for the well-being of the people of the United States and potentially the entire world.

Acknowledgment

Rebecca Moritz, University of Wisconsin Select Agent Alternate Responsible Official and Institutional Contact for Dual Use Research for her assistance in reviewing the chapter and her dedication to the UW-Madison Select Agent Program.

References

[1] Florey HW. The use of micro-organisms for therapeutic purposes. Yale J Biol Med 1946;19:101–18.

[2] Fleming A. History and development of penicillin Fleming A, editor. Penicillin: its practical application. Philadelphia, PA: The Blakiston Company; 1946.

[3] Shatz A, Bugie E, Waksman SA. Streptomycin, a substance exhibiting antibiotic activity against Gram-positive and Gram-negative bacteria. Proc Soc Exp Biol Med 1944;57:244–8.

[4] Henkel RD, Miller T, Weyant RS. Monitoring select agent theft, loss and release reports in the United States – 2004–2010. Appl Biosafety 2012;17:171–80.

[5] Committee on Recombinant DNA Molecules Assembly of Life Sciences National Research Council Potential hazards of recombinant DNA molecules. Proc Nat Acad Sci USA 1974;71:2593–4.

[6] Berg P, Baltimore D, Brenner S, Roblin III RO, Singer MF. Summary statement of the Asilomar conference on recombinant DNA molecules. Proc Nat Acad Sci USA 1975;72:1981–4.

[7] Convention on the prohibition of the development, production and stockpiling of bacteriological (biological) and toxin weapons and on their destruction. 10 April 1972. Available from: <http://www.un.org/disarmament/WMD/Bio/> [accessed 16/8/14].

[8] The Biological Weapons Anti-Terrorism Act of 1989; 18 USC sec. 175.

[9] The Antiterrorism and Effective Death Penalty Act of 1996. April 24, 42 U.S.C.

[10] Uniting and Strengthening America by Providing Appropriate Tools Required to Intercept and Obstruct Terrorism (USA PATRIOT Act) Act of 2001. (Public Law 107-56, October 26, 2001).

[11] U.S. Congress. Public Health Security and Bioterrorism Preparedness and Response Act of 2002. P.L. 107-188. 42 U.S.C. 243, June 12. Available from: <http://tis.eh.doe.gov/biosafety/library/PL107-188.pdf> [accessed 16/8/14].

[12] Select Agent Rules. DHHS, Public Health 42 CFR Part 73 (Select Agents and Toxins); USDA, Agriculture (Plants) 7 CFR Part 331 (Possession, Use and Transfer of Select Agents and Toxins); USDA, Animals and Animal Products 9 CFR Part 121 (Possession and Transfer of Select Agents and Toxins).

[13] Wilson DE, Chosewood LC, editors. Biosafety in Microbiological and Biomedical Laboratories (BMBL) (5th ed.). : U.S. Department of Health and Human Services; 2009. Section VI; pp. 105–113 HHS Publication No. (CDC) 21-1112. Available from: http://www.cdc.gov/biosafety/publications/bmbl5/.

[14] Committee on Research Standards and Practices to Prevent the Destructive Application of Biotechnology Biotechnology research in an age of terrorism. National Research Council; 2004. Available from: <http://www.nap.edu/catalog/10827.html>.

[15] National Science Advisory Board for Biosecurity. Proposed framework for the oversight of dual use life sciences research: strategies for minimizing the potential misuse of research information. Available from: <http://osp.od.nih.gov/office-biotechnology-activities/biosecurity/pdf/Framework%20dortransmittal%200807_Sept07.pdf>; 2007 [accessed 16/8/14].

[16] NIH Guidelines for research involving recombinant of synthetic nucleic acid molecules. Available from: <http://osp.od.nih.gov/sites/default/files/NIH_Guidelines_0.pdf>; 1976 (revised 2013) [accessed 16/8/14].

[17] Richardson JH, Barkley WE, editors. Biosafety in microbiological and biomedical laboratories (5th ed.). Washington, DC, USA: Department of Health and Human Services; 1984. HHS Publication No. (CDC) 86-8395.

[18] U.S. Government. United States Government policy for oversight of life sciences dual use research of concern. Available from: <http://osp.od.nih.gov/office-biotechnology-activities>; 2012 [accessed 27/8/14].

[19] U.S. Government United States Government policy for oversight of life sciences dual use research of concern. Fed Regist 2013;78(36):12369–72.

[20] U.S. Government. A Framework for Guiding U.S. Department of Health and Human Services Funding Decisions about Research Proposals with the Potential for Generating Highly Pathogenic Avian Influenza H5N1 Viruses that are Transmissible among Mammals by Respiratory Droplets. Available from: <http://www.phe.gov/s3/dualuse/Documents/funding-hpai-h5n1.pdf>; 2013 [accessed 27/8/14].

[21] *United States Government Policy for Institutional Oversight of Life Sciences Dual Use Research of Concern*. United States Government, 2015. Available from: <http://www.phe.gov/s3/dualuse/Pages/default.aspx>.

[22] AAAS, AAU, APLU, FBI. Bridging science and security for biological research: a discussion about dual use review and oversight at research institutions. Available from: <http://www.aaas.org/search/gss>; 2012 [accessed 27/8/14].

[23] Imai M, Watanabe T, Hatta M, Das SC, Ozawa M, Shinya K, et al. Experimental adaptation of an influenza H5 HA confers respiratory droplet transmission to a reassortant H5 HA/H1N1 virus in ferrets. Nature 2012;486:420–8.

[24] Herfst S, Schrauwen EJA, Linster M, Chutinimitkul S, de Wit E, Munster VJ, et al. Airborne transmission of influenza A/H5N1 virus between ferrets. Science 2012;336:1534–41.

[25] World Health Organization. Public health, influenza experts agree H5N1 research critical, but extend delay. Available from: <http://www.who.int/mediacentre/multimedia/h5n1pressbriefings/en/>; 2012 [accessed 21/10/14].

[26] University of Wisconsin-Madison new DURC Process as of 11/2014 to satisfy the 2015 DURC Policy requirement of a formal institutional IRE – personal communication – Rebecca Moritz, Alternate Responsible Official: The three-person IBC subcommittee will be replaced by three voting members of the IBC as before with appropriate scientific expertise and two ex-officio non-voting members. The latter two are the Alternate Responsible Official who will act as the ICDUR and the chair of the IBC who will be responsible for communication to Office of Biosafety, Responsible Official, and university administration as needed.

[27] National Science Advisory Board for Biosecurity. Statement of the NSABB. March 29–30, 2012 Meeting of the National Science Advisory Board for Biosecurity to Review Revised Manuscripts of Transmissibility of A/H5N1 Influenza Virus. Available from: <http://osp.od.nih.gov/office-biotechnology-activities>; 2012 [accessed 21/10/14].

[28] Council of Government Relations. Federal regulatory changes since 1991. Available from: <http://www.cogr.edu/viewDoc.cfm?DocID=151793> [accessed 17/11/14].

[29] AAU, APLU, COGR. Regulatory and Financial Reform of Federal Research Policy. Recommendations to the NRC Committee on Research Universities. 21 January 2011.

[30] Blaine JW. Establishing a national biological laboratory safety and security monitoring program. Biosecur Bioterror 2012;10:396–400.

[31] Clinton HR. Remarks at the 7th biological and toxin weapons convention review conference. Available from: <http://iipdigital.usembassy.gov/st/english/texttrans/2011/12/2011 1207104803su0.7202352.html#ixzz3Xg1CTUzM>; 2011 [accessed 31/4/15].

[32] AAAS, AAU, APLU, FBI. Bridging science and security for biological research: International Science and Security. Available from: <http://www.aaas.org/search/gss>; 2013 [accessed 17/11/14].

[33] World Health Organization. Biorisk management: laboratory biosecurity guidance. Available from: <http://www.who.int/csr/resources/publications/biosafety/WHO_CDS_EPR_2006_6/en/>; 2006 [accessed 17/11/14].

Emerging technologies and bio-threats

7

David M. White, Charles E. Lewis and Jens H. Kuhn

Chapter Outline

The role of the Institutional Biosafety Committee: risk assessment and protection of the community through (dis)approval of proposed scientific activities

Scientists voluntarily convened in 1973 (Gordon Research Conference on Nucleic Acids) and 1975 (Asilomar Conference on Recombinant DNA Molecules) to discuss societal concerns surrounding the newly emerging recombinant DNA (rDNA) technology. As a result of these conferences, guidelines, and recommendations were drafted on how to perform rDNA experiments safely without endangering the laboratory worker, the environment, or society as a whole. Institutional Biosafety Committees (IBCs), soon mandated by the National Institutes of Health (NIH) for all entities receiving NIH funding, became a standard at US research institutions as a direct consequence of these recommendations. However, at some institutions, their role has expanded beyond the evaluation of rDNA technology. This chapter will summarize the current challenges that emerging biological agents and emerging technologies present to IBCs in experimental risk assessment and communication, and outline a path forward on how to address them.

In a testament to the foresight of the attendees of the 1975 Asilomar Conference, the text below applies to the challenges of assessing emerging technologies (and agents) almost as well today as it did nearly 40 years ago (emphasis added):

> *This meeting was organized to review scientific progress in research on recombinant DNA molecules and to discuss appropriate ways to deal with the*

Ensuring National Biosecurity. DOI: http://dx.doi.org/10.1016/B978-0-12-801885-9.00007-X

*potential biohazards of this work. Impressive scientific achievements have already
been made in this field and these techniques have a remarkable potential for
furthering our understanding of fundamental biochemical processes in pro-
and eukaryotic cells. The use of recombinant DNA methodology promises to
revolutionize the practice of molecular biology.* **While there has as yet been no
practical application of the new techniques, there is every reason to believe that
they will have significant practical utility in the future.**

*Of particular concern to the participants at the meeting was the issue of
whether the pause in certain aspects of research in this area, called for by the
Committee on Recombinant DNA Molecules of the National Academy of Sciences,
U.S.A. in the letter published in July, 1974, should end; and, if so, how the scientific
work could be undertaken with minimal risks to workers in laboratories, to the
public at large and to the animal and plant species sharing our ecosystems.*

*The new techniques, which permit combination of genetic information from
very different organisms, place us in an area of biology with many unknowns. Even
in the present, more limited conduct of research in this field, the evaluation of
potential biohazards has proved to be extremely difficult. It is this ignorance that
has compelled us to conclude that it would be wise to exercise considerable caution
in performing this research. Nevertheless, the participants at the Conference
agreed that most of the work on construction of recombinant DNA molecules
should proceed provided that appropriate safeguards, principally biological and
physical barriers adequate to contain the newly created organisms, are employed.
Moreover, the standards of protection should be greater at the beginning and
modified as improvements in the methodology occur and assessments of the risks
change. Furthermore, it was agreed that there are certain experiments in which
the potential risks are of such a serious nature that they ought not to be done with
presently available containment facilities. In the longer term serious problems may
arise in the large-scale application of this methodology in industry, medicine and
agriculture.* **But it was also recognized that future research and experience may
show that many of the potential biohazards are less serious and/or less probable
than we now suspect.**

*(quoted from the Summary Statement of the Asilomar
Conference on Recombinant DNA Molecules [1])*

At its most basic level, the role of the IBC is to evaluate and ensure the safety of
proposed research or scientific activities to both workers and the larger environment
where their institutions reside. After the groundwork laid by the pioneers in this
field, such risk assessments are sometimes considered routine, and are usually a part
of the regulatory and oversight requirements faced by most institutions. However,
due to the variable and (occasionally) unpredictable nature of emerging agents and
technologies, highly specialized and/or cross-discipline knowledge (e.g., knowledge
of niche subspecialties of chemistry, physics, synthetic biology, or even engineering)
is sometimes required to adequately evaluate incoming research proposals. Facing
proposed research outside their specialized knowledge, IBC members may err on the
side of caution, thereby potentially halting or preventing highly promising research.
Alternatively, the committee may rely too heavily on the submitter's judgment of his/
her proposal and approve risky activities that should receive a deeper level of scrutiny.
In such a case, the temptation to consult *ad hoc* reviewers will be great and should

be encouraged. Such consultation can be an excellent solution to a real or perceived lack of expertise on an emergent technological or biological hazard. Unfortunately, at many institutions (especially smaller or isolated institutions), there is unlikely to be an in-house, objective third party that can effectively evaluate highly specialized research proposals on novel technologies or emerging biological threats. Distinct research groups at the same or different locations that are engaging in research with the same emerging agents or technologies often collaborate or compete, raising questions of objectivity – even if only in perception. It was noted without irony that the moratorium on rDNA experiments that culminated in the 1975 Asilomar Conference was "vulnerable to portrayal as the fox set to guard the chicken coop" [2].

Additionally, seemingly similar agents or technologies can respond to manipulation differently in some significant ways, raising questions of the applicability of objectively applied but off-topic expertise. Also, while there is no restriction on bringing in expertise from outside the institution, logistics can be problematic. At a minimum, outside reviewers should be aware of their duty to confidentiality (if not required to sign a statement to that effect), should be vetted for conflicts of interest similar to internal reviewers, and be available and committed to the full term of adjudication in case the matter becomes more complex over time. Geographically isolated research institutions may need to sponsor travel, but will likely handle the review process using technological approaches such as secure email or conference calls. While an IBC cannot cede its review process or judgment to the proposal's authors, it must at the same time be aware of the benefits as well as the potential pitfalls of consulting with external reviewers.

In a seminal paper, Dunning and Kruger outlined a "dual burden" of the unskilled: "Not only do these people reach erroneous conclusions and make unfortunate choices, but their incompetence robs them of the metacognitive ability to realize it." This became known as the Dunning–Kruger effect. However, a relevant corollary to the noted effect was that intelligent people often under-rated their abilities in relation to their peers [3]. This lack of confidence in one's own abilities could lead individual IBC members (or entire committees) to inappropriately cede their judgment(s) to third parties that may be no more capable than the committee members themselves. It must be remembered that science (mostly) progresses slowly, in incremental steps aimed at fulfilling the scientific method; as such, novel topics are generally going to be accessible to a scientifically educated committee member with a willingness to undertake some research and analysis in the performance of due diligence. At the inception of the rDNA Advisory Committees (RACs) and IBCs, members had to acquire a basic understanding of knowledge current researchers now take for granted in assessing recombinant or synthetic nucleic acid research proposals. Today, members must be reminded of those approaches when examining complex or novel studies. An example may be a proposal on intercalating drugs that drive RNA viruses to "error-extinction" by disrupting the quasispecies population beyond its ability to carry genetic informational identity into the next generation [4]. A basic understanding of the nature of genetic identity and the redundant genetic code, coupled with research on the nature of replication fidelity of RNA viruses, will make the concept of disruption of information-carrying capacity accessible to anyone with a strong and relatively current biological educational foundation. Further, an understanding of (or research concerning) the nature of viral genomes and viral replication versus the intracellular sequestration of those activities will allow the committee to assess the risk of the proposed drug to the workers developing it.

Several factors will be relevant to this risk assessment. For instance, whereas a high-molecular-weight drug (actively delivered to the cytoplasm) that binds to RNA will likely bind to all RNA, it will likely be excluded from the nucleus and endoplasmic reticulum (ER) by normal cellular processes, and will therefore intercalate preferentially into viral mRNAs, rather than cellular mRNAs. Conversely, a small-molecule drug that crosses the cellular membrane passively will likely cross the nuclear/ER membrane as well, and therefore might be considered more hazardous to cellular processes in the absence of other information. While not a guarantee of absolute safety, the process exemplified above is a part of a viable and robust assessment of the risk of a proposed scientific activity, and meets the duty of an IBC in protecting workers and the public.

Ideally, IBC evaluation of emerging technologies will be based on scientifically acquired knowledge alone. The transgenic research community has already faced difficulty in this regard, with examinations of public attitudes about genetically modified organisms (GMOs) finding that "(t)here are only weak correlations between knowledge and attitudes and knowledge and acceptance of GMOs, and a strong correlation between attitudes and acceptance" [5]. Other critical reviews declared that GMOs are "the new food safety concern which, despite the intense reactions from Non-Governmental Organizations (NGOs) and consumer organizations, have entered our lives with inadequate legislative measures to protect consumers from their consumption" [6]. Efforts to educate the public and reframe discussions away from uncertainty about the unknown to the societal implications of the work, as well as attempting to instill a fundamental understanding of transgenic organisms as the logical extension of rDNA technology, have proven difficult. Polls continue to show a tragic mistrust of the scientific community on this issue. This ongoing conversational confusion has been propagated into the realm of synthetic biology ("The danger is not just bioterror, but 'bioerror'" [7]), "gain-of-function" research on infectious agents ("The same people continue to make the same arguments, and some scientists 'feel like they're treading water'" [8]), and the use of recombinant proteins as growth or production promoters in food and fiber animals ("Cynical activists have unfairly stigmatized a scientifically proven product that has consistently delivered economic and environmental benefits to dairy farmers and consumers" [9]). The IBC can and should play a critical role in communicating the fundamental scientific aspects of such novel work, the risk assessment process, and the essential truth that we, as a society, are safer, more adaptable, and more responsive to natural or man-made threats through increasing our knowledge of the world around us. Simultaneously, it is our thesis that we are made less safe, less adaptable, and less responsive by closing off viable and valuable avenues of research. IBCs have, and should continue to have, the "Power of No"; however, they should use that power only when their decision is based in rational and scientific reasoning absent from partisan or demagogic influence.

When investigators propose to work with novel agents or biotechnologies, authors and reviewers must conduct their discussions openly and pursue direct communication with all stakeholders. Often, such transparency both raises and resolves legitimate concerns. A high standard of communication is necessary, and authors must try to anticipate the needs of reviewers and submit relevant information in a way that allows a scientifically educated reviewer who is not an expert in a specific area to fully assess possible risks of the proposed work. Many laboratory activities are seen as routine, especially by investigators, based on long experience with them or knowledge about their implications to the natural world. However, it must be remembered that these routine activities were once novel technologies that may have created anxiety in reviewers or the public, and might still do so today. For example, in the early days of rDNA manipulations in bacteria, the inclusion of a β-lactamase-encoding gene in plasmids to achieve ampicillin resistance for selection of transformants was seen as highly controversial [10]. That controversy disappeared once it was highlighted that β-lactamases are widespread in nature, and that this fact had been known since 1947 [11]. While a β-lactamase gene could (theoretically) "escape" from a laboratory, it is logical that such a release is in all likelihood inconsequential based on its natural prevalence. The standard for communication of such facts was higher when this technology was novel than it is in current times, when plasmids containing β-lactamase genes are present in commercially available, unregulated kits.

On the other hand, novel uses of biological agents as components of emerging technologies may be proposed by scientists who know relatively little about these agents and may not be able to provide adequate information for review. For example, using a filamentous virus as a molecular engine to aid in the creation of nanomaterials [12] may put a physicist in the position of submitting to an IBC for the first time in his/her career. In this situation, an IBC can – and we would argue should – provide guidance on biological safety to the submitter, such as suggesting alternative organisms consistent with the Asilomar conference findings to limit the risk of rDNA experiments through "the use of biological barriers [such as] non-transmissible and equally fastidious vectors (plasmids, bacteriophages or other viruses) able to grow only in specified hosts" [1]. Such guidance should be provided to facilitate the review process, but the committee must be open to an ongoing discussion of how and why the author designed the experiments as submitted. For example, expert IBC members might suggest the use of tobacco mosaic virus instead of a virus pathogenic to humans or animals as the molecular scaffolding for nanowire assembly. The researcher may accept this recommendation or respond with biological details of the originally proposed virus that are integral to the hypothesized success of the experiments. However this conversation resolves, in this situation, the IBC's follow-up role of ensuring that working conditions mitigate risks and meet safety requirements may be more important than the proposal review role.

IBCs cannot consider any laboratory activities with live pathogenic organisms or viruses (or recombinant nucleic acids for those IBCs with an originalist scope) to be beyond review, even if the organisms are attenuated in healthy, adult humans (the *BMBL* standard for susceptible hosts in biosafety level classification [13]), or thought incapable of causing disease. When a veterinarian in the western United States hears hoof beats, it is not wise for him/her to search for a zebra. However, that same veterinarian must be aware of the existence of zebras (or adult humans that aren't healthy, or children, etc.), and be open to the possibility that there may be one in the vicinity. Tragic failures of biosafety have been extensively reported, some due to failures of work practice [14] and some due to unexpected biological interactions [15]. Thus, no work with live pathogenic organisms is (or should be) beyond IBC review. While rare, exposures, infections, and deaths in the setting of laboratory research may be prevented with strict adherence to general biosafety practices along the lines of Standard Precautions in healthcare delivery. Recent biocontainment failures might also have been avoided with adherence to well-established laboratory protocols requiring inactivation of biological agents before transfer [16] or implementing proper inventory/ cataloging provisions for pathogen collections [17]. The US government's response to these biocontainment failures will likely reduce the prospect of any further issues similar to those that occurred, at least in the near future. However, this response does not include a general mandate to anticipate and solve tomorrow's problems using generally applicable tools and technologies available today [18,19]. At a minimum, it must be remembered that even research largely considered low-risk (e.g., with exempted organisms such as *Escherichia coli* K-12 vector systems) must be evaluated for risk by the researcher. Most IBCs will require the researcher to document that review to the IBC, either before work begins, or (more commonly) concurrent with the work through an expedited review process, to ensure that the committee concurs with the researcher's opinion. Very few IBCs will not review such experiments at all.

Experiments with emerging agents and/or technologies are, by definition, situations for which very little experience, previous critical examination, and/or previous review exists. It took many years to understand that the "bad air" that spread "mal-aria" in the Americas is, in reality, mosquito vectors of specific species, with defined biting behaviors and competencies for transmission of the actual malaria-causing plasmodia. In more recent times, transmission puzzles still abound; as one example, the newly discovered porcine epidemic diarrhea virus (PEDV) is spreading like magical "bad air" through the swine industry, which has a reputation for practicing a high level of agricultural biosecurity (i.e. pathogen exclusion techniques). Using some traditional and some modern approaches, PEDV was rapidly discovered to be an alphacoronavirus, suggesting transmission modes similar to other porcine alphacoronaviruses (e.g. transmissible gastroenteritis virus [TGEV] and its predilection towards spreading via aerosol or contact exposure). However, as of this writing, measures effective against TGEV have proven to be of limited efficacy against PEDV spread [20].

Facing this kind of uncertainty, IBC members might be tempted to take a very stringent approach to review proposals including isolates of emerging biological agents. However, such an approach could unnecessarily hinder valuable research that could be conducted in an attempt to understand and control disease outbreaks. As PEDV became present in an increasing number of states, at what point does it transition from being an emerging pathogen to an endemic/established virus? Should each IBC of each PEDV research institute be made aware of the local epizootiology and change its risk assessment accordingly, and perhaps dynamically? Each committee would have to come to its own individual and collective decision on such matters, but should be reminded that the 1975 Asilomar Conference attendees believed that "the standards of protection should be greater at the beginning [of our understanding of agents/technologies] and modified as improvements in the methodology occur and assessments of the risks change" [1].

A prime example of the dilemma of IBCs is so-called "gain-of-function" research, i.e. research that aims to genetically modify pathogens with specific biological properties to address pressing scientific issues, such as whether a pathogen could easily mutate into a more aggressive or more transmissible form in nature. Two similar experiments, both successful in making a highly pathogenic avian influenza A (HPAI) H5N1 virus more transmissible in an animal model, led to considerable debate about the danger that such a recombinant virus could escape the laboratory. In the end, the debate resulted in a moratorium on certain experiments with H5N1 until the issue was fully debated and a consensus was achieved. For such a debate to be constructive, it must be held to high standards of rationality and scientific theory, free of demagoguery. A wide variety of expertise is needed, encompassing experience with influenza A viruses, animal models for influenza, and influenza A virus biocontainment elements. The moratorium on certain types of rDNA experiments called for in 1974 [21] limited future work. In an unprecedented recommendation, a 2011 moratorium restricted the publication of the data from existing experiments for 6 months, and was extended to restrict work for approximately a year [22–24]. The unavailability of the data due to the moratorium on publication of experimental results, despite the fact that the experiments in question had been reviewed and approved by granting agencies, institutional IBCs, and collaborating researchers, led to a significant amount of uncertainty in the influenza research community with regard to which experiments are permitted and/or publishable. More importantly, the absence of publications at the time made it impossible for the larger scientific community to examine the accumulated data and form a rational assessment of the risks proposed by these influenza A virus experiments. After such an examination by expert panels, the (limits of the) mutagenic potential of the influenza A virus [25], the biotechnological controls designed into the recombinant influenza A viruses [24], and the biocontainment level used at the institutions performing the work suggest that the moratorium on both research and publication was more likely driven by the scientific community's attempts to be responsive to a worried public rather than by substantive concerns among specialists. A similar moratorium was recently instituted based on the emergence of Middle East respiratory syndrome coronavirus (MERS-CoV), and the net was cast broadly to include MERS-CoV, severe acute respiratory syndrome coronavirus (SARS-CoV), and HPAI [26].

Each IBC member should ask the question of how he/she would evaluate research such as these influenza A virus "gain-of-function" experiments. Imai et al. [24] used biological controls in constructing their mutant H5N1 influenza A viruses, consistent with the principles outlined at Asilomar, whereas Herfst et al. [23] used mutagenesis and directed evolution in the generation of their ferret-transmissible H5N1 strain. Studies from the Kawaoka laboratory showed that predictable changes in the genome of influenza A virus could extend aerosol transmission to a mammalian model of human infection; similar studies from the Fouchier laboratory used known but more unpredictable methods to find that multiple sets of mutations might achieve the same biological result. These different approaches lead to different conclusions and answer different questions, but does that difference impact the IBC's risk assessment? Is there a rational distinction between the two resultant viruses that would argue for a higher containment level for one or the other?

In December 2011, the National Science Advisory Board for Biosecurity (NSABB) recommended that details from the two manuscripts that provided information on mutations which allowed H5N1 influenza A virus to become transmissible in mammals be redacted [27]. Over the course of the next 6 months, continuing deliberations led to the full and complete publication of those manuscripts [23,24] while continuing a moratorium on similar work for another 6 months. While such work was eventually given permission to restart, it is again now subject to another moratorium, this time in funding. The Gordon conference in 1973 and the Asilomar conference in 1975 openly discussed the emerging technology of recombinant modifications, and brought together accomplished experts to chart a path forward that was scientifically ambitious at the same time that it called for safety and critical review of proposed experiments. The decision to call for a redaction of data on mutations that had been theorized or shown to condition host range in influenza A viruses as far back as 1993 [28,29] cannot be seen solely as a rational response to the scientific process, or the data it generated. We support the eventual publication of these data, and it is our belief that the ongoing NSABB deliberations concerning the influenza A virus "gain-of-function" experiments serve the public interest and scientific community's long-term interests. Technical experts from multiple disciplines ultimately made the decision that the data should be published largely unchanged, with recommendations that follow-up research should continue (with a few modifications). Similarly, future experiments and/or technologies should be subjected to rational discourse that both limits the risks of the work through application of technological solutions (e.g. Asilomar's "biological and physical barriers") and facilitates the research in an effort to advance our understanding of the world around us.

Most importantly, such rational discourse should include the public in the case of controversial experiments, and should be conducted prior to research being approved or initiated. It is clear that "controversy" cannot be reliably anticipated, as the experiments in question could be seen as a simple application of existing data about relevant mutations to a virus of concern to public health specialists across the world, which was reviewed and approved at many different levels [30]. However, what became clear is that a larger risk communication initiative is needed, and more robust, formal, and comprehensive review processes are needed to fulfill our duty to both protect and address the concerns of the public at large that funds our research. Several review structures have been proposed, and some of them focus on the IBC as the entity that should perform assessments of controversial research [31]. It would appear that the current entities slated to complete these reviews are the NSABB and the National Research Council of the National Academy of Sciences [26], but each IBC would be well advised to watch for additional guidance on this subject that may direct them to conduct specific review processes or address specific foci of concern.

Novel biological agents can be (and have been [32–34]) rapidly characterized with existing, broadly applicable technologies. Such characterization can serve as a basis to guide risk analyses of further examinations, and provide early guidance on public/animal/plant health control measures to attempt to limit the spread of emerging diseases. While not perfect (see PEDV above), such guidance is often useful to both the scientific community as well as the public (e.g. SARS and "social distancing" [35,36]). Such control measures can have a secondary, calming effect on the fear associated with an uncertain field. When it is realized that the same measures employed to fight the spread of "known" diseases such as seasonal influenza are effective against emerging diseases, intelligent but scientifically less educated people may intuit the similarity of biological agents in general, and the (relatively) few methods through which they can be spread from infected to naïve individuals. By linking known diseases with the unknown scourges of "emerging biothreats", we propose that both IBC members and the public will take comfort from the application of knowledge and technologies that are widely available, actualized, and based in the immutable rules of Pasteur's Germ Theory.

True scientists are professional skeptics, but the laundry list of new, emerging, or even as-yet-unknown agents and/or technologies that could threaten humans is large and growing. Scientists must keep an open mind about the possibility of arsenic-based bacteria being able to kill humans ([37], even if methodological problems bring the discovery into question [38,39]); randomly generated synthetic genes that spread through (and impair) whole ecosystems [6,40]; syndromic conditions for which scientists can only establish causation through the use of a modified set of Koch's postulates [41–43]; biologically active nanomaterials [44]; or other truly novel biological agents or technologies that present a threat to our community and, perhaps, our society. However, scientists must not allow these valid and useful precedents (and the thought experiments they engender) to unduly influence the fundamental understanding of probabilities, relative risk, and rational assessment of data that is an essential part of the scientific enterprise and (of course) the IBC review process.

IBCs play a vital role in the pathogen discovery, response, and characterization process. By providing a formal and (ideally) objective assessment of risk, they can facilitate dissemination of existing data that were used in their risk assessment, as well as guide future research in a way that protects laboratory workers, the public, and the environment. IBCs must fully evaluate the safety implications of emerging agents and technologies to fulfill societal expectations of review that derive directly from the promises of the 1975 Asilomar Conference. However, IBC members must also recognize that we are ultimately safer with increased knowledge and tools for response, and facilitate research and technology development as an active initiative to combat ignorance of the unknown. These conflicting priorities necessitate a wide knowledge base that will generate relevant questions that attempt to fully evaluate crucial elements of proposed work, and allow an effective and diligent evaluation without ceding judgment and duty(ies) to the same experts that are being evaluated/regulated. Creative and adaptive thinking – in the context of rational examination of known facts and technological paradigms – is necessary to facilitate research in a society that increasingly mistrusts the scientific enterprise [45,46]. This need is especially stark when that society is faced with truly novel agents or technologies. A full, complete, and robust risk assessment should both ensure rational limits on activities and provide reasonable assurances of safety in the face of rapidly changing scientific conditions or knowledge.

Risk assessment of emerging technologies or biothreats

Researchers, when faced with an "unknown" usually ask the question "what is known about its closest relatives and/or similar technologies?" If a scientist submitting a proposal to an IBC has not provided sufficient information on their protocol/risk assessment, the scientist should be encouraged to submit additional supporting details for review (and the process may well be iterative). At a minimum, the scientist must expressly state the limits of current knowledge about the object of proposed research, fully elucidate hypotheses, and fully describe both experimental approaches and the biosafety controls (e.g. administrative, facilities, work practices, and PPE/safety equipment) to be utilized for the work. The scientist must submit both necessary and sufficient information such that a knowledgeable reviewer can fully assess the work, even though the scientist may desire to limit the submitted information in an effort to decrease their work burden when submitting protocols for review. Guidance documents and committee policy statements can be created to give submitters an idea of the types and depth of information needed for adequate review.

IBCs should evaluate the provided information in the context of its existing knowledge base without falling prey to assumptions that similarities of the "unknown" agent or technologies to known agents or technologies are necessarily predictive of risks. For instance, coronavirus research was a focused subspecialty until the discovery of SARS-CoV – and although much was known about coronaviruses by the time of SARS-CoV's emergence, the lethality of SARS-CoV for humans could not have been predicted, as coronaviruses highly virulent for humans were not known at the time. It is important to keep such examples in mind during a robust risk assessment process concerning novel agents or technologies.

At the same time, IBCs also must ensure that they are not overestimating risk because too little information is known. Classifying an uncharacterized agent at a high biosafety level based only on the fact that it is uncharacterized is rarely helpful. The possible pandemic of SARS-CoV was (in part) blunted because many studies commenced in laboratories around the world immediately after the virus was isolated, despite the fact that it was a completely novel agent. In the unlikely event that no background or comparative information exists that can be outlined in a research proposal to the IBC, the IBC can suggest a course of pre-experimental actions based on general biological principles to generate sufficient data for consideration. First and foremost, the early events in the SARS investigations instruct us that education, training, and strict adherence to basic laboratory containment practices are essential in the investigation of novel agents, and could have prevented fatal breaches in standard laboratory practice [47]. The original Asilomar attendees anticipated this need and noted that "(p)articularly important is strict adherence to good microbiological practices which, to a large measure, can limit the escape of organisms from the experimental situation and thereby increase the safety of the operation. Consequently, education and training of all personnel involved in the experiments is essential to the effectiveness of all containment measures" [1].

Broadly reactive primer sets or group-reactive antibodies can be used to rapidly characterize novel viruses or bacteria, or (at a minimum) to rule out similarity to tested groups. Multiple lines of testing should be initiated, examining as many essential (e.g. physical, chemical, and biological) properties of the novel agent as possible. Electron microscopy to detect morphological cues for classification, nucleic acid detection (e.g. group-specific polymerase chain reactions, targeted or random probe hybridization microarrays, etc.) followed by genome sequencing, protein identification (e.g. group-specific antibody testing or sequencing using mass spectrometry), and growth in permissive systems (e.g. intracranial inoculation of suckling mouse brain, cell panels) should all be attempted in the early stages of identification of novel agents. If a successful culture system is found, basic risk assessments can be conducted examining growth in immortalized human-origin cell lines, or perhaps even human organ-system models (e.g., organs-on-chips, 3D cultures) or primary cells (if available).

Similarly, novel technologies must be related to their source in existing, well-characterized technologies. rDNA technologies were examined in the context of the prescient work on horizontal gene transfer conducted in bacteria before the structure of DNA was known [48], and recombinant organisms and synthetic biology derive from (and can draw from the conclusions of) those early technologies. Gain-of-function experiments exist on the backdrop of the parental biological agents; novel diagnostic technologies can be characterized by sensitivity, specificity, and cost; and even the probability of the resurrection of a *Tyrannosaurus rex* can be anticipated (even if only by Hollywood), calculated, assessed, (perhaps) mitigated, and at a minimum communicated to an interested public that likely would fund the research.

Consistent with the original 1975 Asilomar Conference report [1], "(e)stimating the risks will be difficult and intuitive at first but this will improve as we acquire additional knowledge; at each stage we shall have to match the potential risk with an appropriate level of containment". Early characterization experiments should be done at a level of containment that may be in excess of what is used after the risk of the experiments is better known or fully studied. For those novel agents that are recognized due to a human disease outbreak, crucial experiments with live agent that demonstrate susceptibility (or resistance) of the "unknown" agent to possible treatments, cross-reactivity with available vaccines, and/or permissiveness of human cells to infection or biological dysfunction should be performed at a containment level commensurate with the information surrounding the novel agent's/toxin's/technology's discovery/emergence. In a practical sense, novel agents that are recognized due to high virulence, or are closely related to highly virulent viruses, should be initially characterized at higher (e.g. BSL-3 or above) biosafety levels or maximum containment. Those agents that are discovered by research into multifactorial syndromes and/or as adventitious agents not specifically linked to a disease in humans, animals, or plants can be dealt with at a lower containment level depending on an assessment of their risk to laboratory workers and/or the environment. This approach, however, is based on the explicit willingness of an IBC to downgrade the biosafety level at which work is initially performed once sufficient data are available that counter or inform the initial risk assessment. In the past, there have been only a few examples of human pathogens that were initially classified at a high biosafety level and then were downgraded. If downgrading is an unlikely event even in the presence of supporting data, it becomes "risky" for a scientist to (for instance) suggest BSL-4 classification of an "unknown" that causes significant disease in humans, as the low number of BSL-4 facilities greatly limits research and therefore the development of possible countermeasures. The Asilomar attendees knew, and expressly stated, that "(t)he means for assessing and balancing risks with appropriate levels of containment will need to be reexamined from time to time" [1]. Biosafety classification should therefore be re-assessed over time, weighing new or developing information that may condition scientific or public opinion about the pros and cons of the work to laboratorian (and overall public) health.

Clearly, the above is a simple treatment of the complexity of a true assessment of risk, lacking the depth that will be required in real-world implementation. For example, risk assessment may change based on a scientist's experience with high-hazard agents, the containment level of the facilities available for experiments, and the institutional commitment to biosafety. The simplicity was deliberate, as a prescriptive set of rules will not allow the flexibility and local control that is an essential strength of the IBC system. The basic approach that smart, rational, and honest people should evaluate research activities smartly, rationally, and with a high degree of intellectual honesty was the core promise of the Asilomar Conference attendees to a nervous public. That approach should be strengthened, continued, and should not be subject to an artificial set of external rules that will ultimately poison the intellectual integrity that each committee member brings to the faithful discharge of their duties.

A possible role of the IBC: efficacy considerations and resource allocation decisions as representatives of the public interest

The mandate of an IBC is to evaluate the safety of proposed research – specifically the biosafety of the proposed activities. A reasonable expansion of that mandate would be to include a consideration of all safety (e.g. chemical, radiological, etc.) to the degree to which committee members have experience with those hazards. Additionally, IBCs are charged with ensuring that the approved work activities are actually performed as stated in the proposal, through a series of follow-up reports, proposal renewals, inspections, and/or "whistle-blower"-type reporting mechanisms. However, it may occur to readers that an additional expansion of the mandate to consider the likelihood of success of proposed research is logical, especially in today's limited funding pool. We would like to argue against this expansion for two reasons. First and foremost, research that proceeded from the conception stage to the proposal submission stage has already been, or is currently being, considered for likelihood of success by the funding agency's subject matter experts. Secondly, novel research should not be stifled by an additional round of review by those that may have less expertise in the specific field being considered. If a scientist has taken the time to submit a protocol for IBC review, it has probably already been funded, unless IBC approval is required before it will be considered for funding. In either case, an IBC should evaluate the researcher's ability to complete the proposed work safely at their institution only.

Imposing opinions about the success of the research onto an already stringent process is a level of review that is at least burdensome and at most unfair to scientists, as well as the larger research enterprise.

Additionally, the voluntary expansion of the IBC mandate to consider all research at institutions (instead of that research that utilizes synthetic or recombinant nucleic acids) has already taxed many committees, requiring expansion of personnel and increased administrative support. To further increase their purview, with little expected benefit, is not a wise use of the intellectual or temporal resources of the committee members. Consideration of efficacy for proposed activities requires a much greater depth of examination than the consideration of safety protocols, containment, and researcher and institutional commitment to safety that make up the current review process.

The role of the IBC: risk communication and mediating the "science conversation" with members of the public

It is essential to remember that one important but under-utilized role of the IBC is to act as a mediator of the conversation between a scientifically minded workforce, and an intelligent but scientifically less educated public. Science communicators have come to know that there is a proportion of minds that cannot be influenced by rational discussion; however, that percentage is much smaller than one might believe after listening to popular news programs or reading a comment section on the internet. Many people are simply in search of information that is both accessible *and* fills in the gaps in their knowledge in a way that increases their understanding of the world around them. One might be hard pressed to find a better definition of the scientific mind than that simple statement concerning curiosity and Feynman's "philosophy of ignorance":

> *It is our responsibility as scientists, knowing the great progress and great value of a satisfactory philosophy of ignorance, the great progress that is the fruit of freedom of thought, to proclaim the value of this freedom, to teach how doubt is not to be feared but welcomed and discussed, and to demand this freedom as our duty to all coming generations.*

If a scientist working in an arbovirology laboratory received a letter from a member of the public asking whether "the AIDS" could be transmitted by mosquitos, should he/she respond, and if so, how? While it might be tempting to disregard the letter, could this writer's fears be alleviated by a basic discussion of the difference between mechanical and biological transmission of viruses by insects, appropriately targeted to the audience's interest and (presumed) educational level? Would it be better to simply say that while their intuition of blood transmission from one mosquito bite to the next was technically possible, there are several facts that argue against such a possibility in the case of HIV/AIDS? Would the latter invite another question, or would it be satisfactory? And, which would be better?

At its most basic level, the role of the IBC is to evaluate the safety of research or scientific activities proposed, to both the workers as well as the larger environment where their institutions reside. While robust examination of those activities by members of the community (other than the public committee members) could present interesting communication opportunities, in the absence of a problem or politically charged topic the community often relies on the IBC to represent their interests. Most IBCs take this duty seriously, and value their public members' input.

Conversely, we have seen failures of this conversation that have been chilling to the performance of research as well as to the larger research enterprise. It must be admitted that both scientists (and the larger scientific community) *and* a potentially suspicious public could play a role in disruptions of the normal flow of science. As scientists, we must look to the "natural experiments" that have been conducted on this issue. The actions that culminated in the Asilomar Conference represent an approach of robust self-examination and self-regulation by the scientific community in response to some real (and, admittedly, some perceived) threats of concern to the larger public. These actions and regulatory structures are largely seen as a success of communication and transparency, especially when contrasted with the regulatory burden placed on researchers engaged in either human or animal experimentation, particularly if the research is hotly debated in the public sphere.

Institutions engaged in human or animal research have a large regulatory burden that requires significant financial and human resources support, which is not present in the implementation of the IBC at most institutions. While important, IBC membership is usually a collateral duty to the research, teaching, or corporate activities that are a committee member's primary tasks. However, Institutional Review Boards (IRBs) and/or Institutional Animal Care and Use Committees (IACUCs) are trending towards hiring dedicated personnel (e.g. IACUC Coordinators and/or Compliance Officers, dedicated or collateral-duty Post-Approval Monitoring personnel). These committees are investing significantly in information technology and/or administrative resources for proposal submission and review, and/or entire "Regulatory Affairs" sections dedicated to compliance with the applicable laws, regulations, and/or guidelines associated with those activities. One could argue that the different subject matter (biosafety in the laboratory vs. intentional impact on living systems) is a direct cause of the difference in regulatory oversight; this argument does not account for the consequences of a failure of oversight or verification in the different systems. Ethically questionable human and/or animal experiments are repugnant and damaging to the morals and/or philosophies of different groups or individuals, but they are largely limited to specific laboratories or experiments and do not generally lead to release of biological agents [49–51]. On the other hand, failures in biosafety and/or laboratory containment could have unpredictable effects on the public (e.g., laboratory release of an agent [52–55]) or the environment [56–58], and could be much more consequential to the population as a whole.

Failure(s) of self-examination and self-regulation have led to a public attitude toward science that has been described as a loss of Aesculapian authority [46,59]. In the past, a physician's order or a researcher's statement of assurance were most often seen as authoritative and definitive statements to calm the fears of the community. That time has clearly (and thankfully) passed, and anyone who counts solely on the authority (supposedly) conveyed by academic or professional degrees lacks a fundamental understanding of the role and position of science in today's society. Should IBCs play an active role in facilitating this conversation, both to facilitate research as well as to repair the damage to our larger scientific integrity caused by historical failures? Or, should IBCs act only as resources to skilled public relations professionals, receive training in science communication before engaging in debate, or be discouraged from any form of official (or unofficial?) communications with the public? Finally, how will those internal rules relate to the committee's duty to be transparent and accessible to the public, codified in Sections IV-B-2-a-(6) and (7) of the NIH Guidelines? Committees are not required to be proactive communicators *per se*, but the lack of communication presents a hazard to the scientific enterprise that should be decreased or eliminated if at all possible.

Each committee must make its own decision on these questions, but several factors must be fully considered. First and foremost, science communication and the conduct of scientific research require very different skill sets, and those accomplished at both are rare and storied individuals. An unskilled but well-intentioned communicator can still cause damage to the relationship between the scientific community and the larger populace, as can a skilled communicator who does not understand the science he or she is communicating.

Further, many antiscience critics are skilled debaters, and will use opportunities of public debate to engage in a demagogic attack on the scientist, his or her work, and the larger scientific enterprise. It has been stated often, by many different groups, that scientists are limited both by their adherence to the truth as well as their professional tenet that one cannot prove a negative, while the other side of the argument is limited by neither consideration [60,61]. Engaging in debate with a flexible and adaptive opposition, in front of a skeptical and fearful public, is not a recipe for successful communication. Indeed, by engaging in such debate, one gives legitimacy and "air time" to the opposition that might open up the possibility for a discussion that could confuse the facts as well as the listening public.

However, an insular scientific enterprise that just wants the concerns of the public to "go away" does not serve humanity in either the short or long term. A robust and diverse IBC that makes rational and faithful decisions that serve the interest of the public should not be afraid to discuss (or even defend) those decisions if necessary. Removing the veil of secrecy that many feel surrounds the scientific community can greatly enhance the research enterprise in many tangible and intangible ways. Imagine the embattled neurobiologist and animal researcher. While committed to their work, these researchers may reevaluate their chosen career paths when an animal rights group targets the researcher's home and/or family [62]. Would the researcher feel more engaged, more effective, and safer if they were viewed differently? A relevant example that presents more broadly applicable health concerns may be the testing of candidate antiviral drugs or vaccines in an outbreak situation (e.g. Ebola virus disease in western Africa, AIDS in southeastern Asia). Are these scientists viewed as people engaged (as respectfully as possible) in troubling but necessary actions in support of public health and a larger, global just and civil society? Can they successfully navigate the moral, ethical, and (perhaps) legal issues inherent in such work? If one believes such work is moral, and perhaps even ethically required, how is that case to be presented?

The "science conversation" is in flux in today's society due to the ever-increasing accessibility to vast amounts of information on the internet, and thought leaders are emerging and setting this stage (e.g. the National Center for Science Education, the American Association for the Advancement of Science, among others) in ways that build upon the traditions created by the original Asilomar Conference. IBCs must wade into this dynamic activity cautiously and with significant forethought for the benefits and drawbacks. If confronted with a communications issue, IBCs should seek resources in the science of communication just as they would seek external resources for the evaluation of novel technologies or agents. The analogy between safety reviews and scientific communications should not be ignored. Executed incorrectly, both may pose significant hazards to the scientific enterprise and can be addressed and mitigated with similar processes.

IBCs play a pivotal role in the review of research involving emerging technologies and bio-threats. This process exists (and must continue to exist) in the context of changing societal norms and expectations, dynamic and creative research, and competing obligations and duties. Committees must embrace and even thrive in the context of the sum (and, occasionally, the synergy) of these diverse interests. While this may be a seemingly impossible task, alternatives pose hazard(s) to the goals of the scientific enterprise, and the larger societal needs it supports. The IBC resides at the nexus of, and must mediate the interactions between, scientists and the public. The trust that resides with these parties must be protected, respected, and appropriately directed toward decreasing risk, increasing knowledge, and facilitating safe science to advance knowledge, safety, and overall societal health.

References

[1] Berg P, Baltimore D, Brenner S, Roblin RO, Singer MF. Summary statement of the Asilomar conference on recombinant DNA molecules. Proc Natl Acad Sci USA 1975;72(6):1981–4.
[2] Wade N. Genetic manipulation: temporary embargo proposed on research. Science 1974;185(4148):332–4.
[3] Kruger J, Dunning D. Unskilled and unaware of it: how difficulties in recognizing one's own incompetence lead to inflated self-assessments. J Pers Soc Psychol 1999;77(6):1121–34.
[4] Domingo E. Quasispecies and the development of new antiviral strategies. Prog Drug Res 2003;60:133–58.
[5] Sorgo A, Ambrozic-Dolinsek J. Knowlege of, attitudes toward, and acceptance of genetically modified organisms among prospective teachers of biology, home economics, and grade school in Slovenia. Biochem Mol Biol Educ 2010;38(3):141–50.
[6] Varzakas TH, Arvanitoyannis IS, Baltas H. The politics and science behind GMO acceptance. Crit Rev Food Sci Nutr 2007;47(4):335–61.
[7] Pollack A. How do you like your genes? Biofabs take orders. New York Times 2007 [September 12, 2007].
[8] Kaiser J. Academy meeting on risky virus studies struggles to find common ground. ScienceInsider [Internet]. Available from: <http://news.sciencemag.org/biology/2014/12/academy-meeting-risky-virus-studies-struggles-find-common-ground>; 2014 [accessed 01.06.2015].

[9] Miller HI. Don't cry over rBST milk. New York Times 2007 [June 29, 2007].

[10] Davies J. Vicious circles: looking back on resistance plasmids. Genetics 1995; 139(4):1465–8.

[11] Bellamy WD, Klimek JW. Some properties of penicillin-resistant staphylococci. J Bacteriol 1948;55(2):153–60.

[12] Zhou K, Zhang J, Wang Q. Site-selective nucleation and controlled growth of gold nanostructures in tobacco mosaic virus nanotubulars. Small 2015;11(21):2505–9.

[13] Wilson DE, Chosewood LC, editors. Biosafety in Microbiological and Biomedical Laboratories (BMBL) (5th ed.). : U.S. Department of Health and Human Services; 2009. HHS Publication No. (CDC) 21-1112. Available from: http://www.cdc.gov/biosafety/publications/bmbl5/.

[14] Centers for Disease Control and Prevention. Fatal Cercopithecine herpesvirus 1 (B virus) infection following a mucocutaneous exposure and interim recommendations for worker protection. MMWR Morb Mortal Wkly Rep 1998;47(49):1073–6. 83.

[15] Centers for Disease Control and Prevention. Fatal laboratory-acquired infection with an attenuated *Yersinia pestis* Strain – Chicago, Illinois, 2009. MMWR Morb Mortal Wkly Rep 2011;60(7):201–5.

[16] Centers for Disease Control and Prevention. Report on the potential exposure to anthrax. Atlanta, GA: CDC; 2014.

[17] Reardon S. 'Forgotten' NIH smallpox virus languishes on death row. Nature 2014;514(7524):544.

[18] Review of CDC anthrax lab incident: hearing before the Subcommittee on Oversight and Investigations, Committee on Energy and Commerce, US House of Representatives (2014).

[19] Written testimony: review of CDC anthrax lab incident: hearing before the Subcommittee on Oversight and Investigations, Committee on Energy and Commerce, US House of Representatives (2014).

[20] Centers for Disease Control and Prevention Fatal Cercopithecine herpesvirus 1 (B virus) infection following a mucocutaneous exposure and interim recommendations for worker protection. Porcine epidemic diarrhea virus (PEDv): American Association of Swine Veterinarians. Available from: <https://www.aasv.org/aasv%20website/Resources/Diseases/PorcineEpidemicDiarrhea.php>; 2015 [accessed 16.03.15].

[21] Berg P, Baltimore D, Boyer HW, Cohen SN, Davis RW, Hogness DS, et al. Letter: potential biohazards of recombinant DNA molecules. Science 1974;185(4148):303.

[22] Fouchier RA, Garcia-Sastre A, Kawaoka Y, Barclay WS, Bouvier NM, Brown IH, et al. Transmission studies resume for avian flu. Science 2013;339(6119):520–1.

[23] Herfst S, Schrauwen EJ, Linster M, Chutinimitkul S, de Wit E, Munster VJ, et al. Airborne transmission of influenza A/H5N1 virus between ferrets. Science 2012;336(6088):1534–41.

[24] Imai M, Watanabe T, Hatta M, Das SC, Ozawa M, Shinya K, et al. Experimental adaptation of an influenza H5 HA confers respiratory droplet transmission to a reassortant H5 HA/H1N1 virus in ferrets. Nature 2012;486(7403):420–8.

[25] Russell CA, Fonville JM, Brown AE, Burke DF, Smith DL, James SL, et al. The potential for respiratory droplet-transmissible A/H5N1 influenza virus to evolve in a mammalian host. Science 2012;336(6088):1541–7.

[26] Office of Science and Technology Policy. Doing Diligence to Assess the Risks and Benefits of Life Sciences Gain-of-Function Research. White House, US Government, 2014. Available from: <https://www.whitehouse.gov/blog/2014/10/17/doing-diligence-assess-risks-and-benefits-life-sciences-gain-function-research>.

[27] National Institutes of Health. Press Statement on the NSABB Review of H5N1 Research. US Department of Health & Human Services, 2011. Available from: <http://www.nih.gov/news-events/news-releases/press-statement-nsabb-review-h5n1-research>.

[28] de Wit E, Fouchier RA. Emerging influenza. J Clin Virol 2008;41(1):1–6.

[29] Subbarao EK, London W, Murphy BR. A single amino acid in the PB2 gene of influenza A virus is a determinant of host range. J Virol 1993;67(4):1761–4.

[30] Fouchier RA, Herfst S, Osterhaus AD. Public health and biosecurity. Restricted data on influenza H5N1 virus transmission. Science 2012;335(6069):662–3.

[31] Faden RR, Karron RA. Public health and biosecurity. The obligation to prevent the next dual-use controversy. Science 2012;335(6070):802–4.

[32] Centers for Disease Control and Prevention Outbreak of Hendra-like virus – Malaysia and Singapore, 1998–1999. MMWR Morb Mortal Wkly Rep 1999;48(13):265–9.

[33] Enserink M. War stories. Science 2013;339(6125):1264–8.

[34] Murray K, Selleck P, Hooper P, Hyatt A, Gould A, Gleeson L, et al. A morbillivirus that caused fatal disease in horses and humans. Science 1995;268(5207):94–7.

[35] Bell DM. World Health Organization Working Group on Prevention of International and Community Transmission of SARS: public health interventions and SARS spread, 2003. Emerg Infect Dis 2004;10(11):1900–6.

[36] Murphy C. The 2003 SARS outbreak: global challenges and innovative infection control measures. Online J Issues Nurs 2006;11(1):6.

[37] Wolfe-Simon F, Switzer Blum J, Kulp TR, Gordon GW, Hoeft SE, Pett-Ridge J, et al. A bacterium that can grow by using arsenic instead of phosphorus. Science 2011;332(6034):1163–6.

[38] Foster PL. Comment on "A bacterium that can grow by using arsenic instead of phosphorus". Science 2011;332(6034):1149. author reply.

[39] Redfield RJ. Comment on "A bacterium that can grow by using arsenic instead of phosphorus". Science 2011;332(6034):1149. author reply.

[40] Chowdhury E, Kuribara H, Hino A, Sultana P, Mikami O, Shimada N, et al. Detection of corn intrinsic and recombinant DNA fragments and Cry1Ab protein in the gastrointestinal contents of pigs fed genetically modified corn Bt11. J Anim Sci 2003;81(10):2546–51.

[41] Fredericks D, Relman DA. Sequence-based identification of microbial pathogens: a reconsideration of Koch's postulates. Clin Microbiol Rev 1996;9(1):18–33.

[42] Falkow S. Molecular Koch's postulates applied to microbial pathogenicity. Rev Infect Dis 1988;10(Suppl. 2):S274–6.

[43] Fouchier RA, Kuiken T, Schutten M, Van Amerongen G, van Doornum GJ, van den Hoogen BG, et al. Aetiology: Koch's postulates fulfilled for SARS virus. Nature 2003;423(6937):240.

[44] Pacheco PM, Le B, White D, Sulchek T. Tunable complement activation by particles with variable size and Fc density. Nano Life 2013;3(2):1341001.

[45] Gauchat G. Politicization of science in the public sphere: a study of public trust in the united states, 1974 to 2010. Am Sociol Rev 2012;77(2):167–87.

[46] Rollin BE. The use and abuse of Aesculapian authority in veterinary medicine. J Am Vet Med Assoc 2002;220(8):1144.

[47] Herman P, Verlinden Y, Breyer D, Cleemput EV, Brochier B, Sneyers M, et al. Biosafety risk assessment of the severe acute respiratory syndrome (SARS) coronavirus and containment measures for the diagnostic and research laboratories. Appl Biosafety 2004;9(3):128–42.

[48] Griffith F. The significance of pneumococcal types. J Hyg 1928;27(2):113–59.

[49] Reverby SM. Tuskegee's truths: rethinking the Tuskegee syphilis study. Chapel Hill, NC: University of North Carolina Press; 2000.

[50] Brandt AM. Racism and research: the case of the Tuskegee Syphilis Study. Hastings Cent Rep 1978;8(6):21–9.

[51] Thomas SB, Quinn SC. The Tuskegee Syphilis Study, 1932 to 1972: implications for HIV education and AIDS risk education programs in the black community. Am J Public Health 1991;81(11):1498–505.
[52] Ferguson NM, Donnelly CA, Anderson RM. Transmission intensity and impact of control policies on the foot and mouth epidemic in Great Britain. Nature 2001;413(6855):542–8.
[53] Gibbens J, Sharpe C, Wilesmith J, Mansley L, Michalopoulou E, Ryan J, et al. Descriptive epidemiology of the 2001 foot-and-mouth disease epidemic in Great Britain: the first five months. Vet Record 2001;149(24):729–43.
[54] Ferguson NM, Donnelly CA, Anderson RM. The foot-and-mouth epidemic in Great Britain: pattern of spread and impact of interventions. Science 2001;292(5519):1155–60.
[55] Shooter RA, Booth CC, Evans D, McDonald JR, Tyrrell DAJ, Williams R. Report of the investigation into the cause of the 1978 Birmingham smallpox occurrence. London: The House of Commons; 1978.
[56] Newsome A, Pech R, Smyth R, Banks P, Dickman C. Potential impacts on Australian native fauna of rabbit calicivirus disease. Biodiversity Group, Environment Australia; 1997.
[57] Mutze G, Cooke B, Alexander P. The initial impact of rabbit hemorrhagic disease on European rabbit populations in South Australia. J Wildl Dis 1998;34(2):221–7.
[58] Pech RP, Hood G. Foxes, rabbits, alternative prey and rabbit calicivirus disease: consequences of a new biological control agent for an outbreaking species in Australia. J Appl Ecol 1998;35(3):434–53.
[59] Siegler M, Osmond H. Aesculapian authority. Hastings Cent Stud 1973;1(2):41–52.
[60] Edwords F. Creation-evolution debates: who's winning them now? Creation/Evolution J 1982;3(2):30–42.
[61] Tanenbaum J. Creation, evolution and indisputable facts. Sci Am 2013;308(1):11.
[62] Miller G. Science and the public. Animal extremists get personal. Science 2007;318(5858):1856–8.

Animal biosafety

Nick Chaplinski

Chapter Outline

Institutional Biosafety Committees (IBCs) are comprised of a chair, community members, biosafety personnel, and individuals with other areas of technical, scientific, and legal expertise. These other experts may include bacteriologists, virologists, veterinarians, or animal model experts, as well as plant researchers. The chair must have a vast understanding of scientific techniques and knowledge. The other members of the committee may also have a wide and varying range of scientific knowledge. Animal models are an essential resource for research related to disease and infection. Animal models can be used in many different ways, including mimicking human infection, replacing the model animal's genes with others – including genes of human origin – to create transgenic animals, or "knocking out" certain genes to determine how loss of function of a given gene affects a particular disease state or the animal's overall health and development.

Many standard animal models sufficiently resemble humans that they can meaningfully predict the effect of infection and disease in humans. Mice are a prime example; mice are surprisingly similar to humans in terms of genetics, physiology, and anatomy – a fact that the average person may not appreciate. Perhaps the greatest factor contributing to the utility of mice as a model of human disease is the fact that 80% of the genes in the mouse genome have a direct counterpart in the human genome [1]. These similarities make it possible to identify and model genetic risk factors in mice that are relevant to human disease. Other animal models can also be useful for more specific research applications. Rats, for instance, are often used for neurological studies as their brains function in a manner similar to humans. The rat and human genomes are also highly similar, and up to 90% of rat genes have direct human counterparts; however, genomic manipulation has proven more difficult in the rat relative to the mouse, and mouse models therefore remain more popular [2].

Ensuring National Biosecurity. DOI: http://dx.doi.org/10.1016/B978-0-12-801885-9.00008-1

Mouse genomes have been more extensively studied to this point, though strides are being made in genetic manipulation of rat genomes, as some symptoms of disease are closer to those of humans in rats than in mice. Non-human primates (NHPs) are the animal models most closely related to humans; the sequences of the chimpanzee and human genomes are 96% identical [3]. However, there are numerous challenges associated with the research use of NHPs, including cost, housing, containment, as well as ethical and public perception issues related to the high intelligence and cognitive function of these species.

In addition to serving as surrogates for humans in experimental research, animal models are useful for studying diseases in natural hosts. These can include diseases that are endemic within a given animal population as well as zoonotic diseases that can be transmitted to human hosts. These include infectious diseases such as foot and mouth disease, Q-fever, herpes B, and brucellosis.

Animals and biosafety

While extremely valuable from a research perspective, the use of animal models also raises specific concerns, many of which relate to biosafety and biosecurity. As one simple but important example, allergies can be of concern for researchers, animal husbandry staff, or anybody else whose job brings them into proximity of the animals. People may be allergic to animal dander, saliva, or urine, or even to the animal bedding. In these instances, personal protective equipment (PPE) such as gowns, sleeves, respirators, or masks may be necessary to prevent allergic reactions to animals in vivariums, satellite animal housing units, or other spaces used for animal procedures. Because allergies represent the most basic level upon which biosafety and biosecurity interventions are built, animal allergy postings, animal biosafety level postings or other relevant information may be provided to inform those entering a certain area that animals are present and provide certain warnings to those that may be affected.

Beyond the risk from biological agents that are shed or excreted after being administered to animals, exposure to metabolites, specifically from drug or chemical treatment, can also be a potential trigger for allergies. While an individual may not have a reaction to a certain agent, drug, or chemical, when excreted in an animal model, these metabolites may cause an adverse reaction. In many cases, these metabolites can be derived from administered toxins of biological origin. Taking this risk into consideration, certain PPEs may be prescribed and standard operating procedures (SOPs) may be modified to address the proper handling and disposal of bedding, as well as how to conduct cage changes. As part of this review process, the IBC may be tasked with determining how the aforementioned issues are to be handled, in conjunction with other chemical or biosafety professionals. In many cases, it may be necessary to develop a cage card system that warns of specific hazards associated with certain agents that were administered to the animal and the risks that may be involved in working with that animal.

It is also important to be aware of the physical challenges of working with animals in a research environment. Animals behave unpredictably, can cause injury with their sharp claws or teeth, and may harbor endemic or administered zoonotic diseases. In the case of large animals, the animals may be physically imposing and able to maim, crush, charge, scratch, bite or throw objects in a research environment. Accordingly, an animal's behavioral disposition or attitude and the degree of personnel experience with that particular species can be an important factor in risk assessments and containment practices as well. It is therefore important to establish effective barriers between workers and animals, and to determine the restraints needed for the animals.

For large animals, one should consider a buddy system in place with experienced large-animal handlers, because of inherently erratic animal behavior. For these reasons, large animal research usually deals with less automation and fewer engineering controls, with heavier reliance on PPE and species-specific knowledge and experience. It is obviously impossible to fit a cow, sheep, or bison into a biosafety cabinet, and in many of these cases – especially in an Animal Biosafety Level-3 Agriculture (ABSL-3Ag) laboratory – the laboratory or research space will serve as the primary containment. With this in mind, special consideration is needed to safely house large research animals, as well as perform research activities and necropsies.

Beyond the fundamental risks related to animal husbandry and manipulation, there are risks associated with the administration of specific agents of biological origin. It is the task of IBCs and its cadre of professions to develop preventative measures and response plans to ensure safety and security by agent and by animal.

Risk assessment, risk groups, and biosafety levels

There are a number of different actions that need to take place before a biological agent is used in an animal model. The initial risk assessment is typically conducted by a biosafety professional, but the entire IBC may provide consultation even at this stage. This risk assessment needs to take into account the type of agent being used, the risk group it belongs to, and different routes of exposure as well as factors such as agent-specific potential for splash or aerosolization. After this initial step, other components of the risk assessment – including review of location, engineering controls, PPE prescriptions and medical surveillance protocols – should be conducted.

Biological agents can be classified into risk groups that start at Risk Group 1 (RG1) and work up to RG4, in increasing order of threat level. RG1 agents are not associated with disease in healthy adult humans. These include a variety of commonly used animal-specific viruses such as adeno-associated viruses. RG2 agents cause diseases in humans that are rarely serious, and for which preventive or therapeutic interventions are often available. Examples include: hepatitis B virus, and many RG2 bacteria easily treated with antibiotics if exposed. RG3 agents are associated with serious or lethal human diseases such as *Mycobacterium tuberculosis* (one of the agents causing tuberculosis), severe acute respiratory syndrome, *Yersinia pestis* (plague), or *Brucella abortus*, for which preventive or therapeutic interventions may be available; this typically means high risk for individual researchers, but lower risk to the general community. Finally, RG4 agents cause serious or lethal human diseases for which preventive or therapeutic interventions are generally not available, creating high levels of risk for both individuals and the outside community [4]. RG4 agents include Ebola virus, Marburg virus and herpes B virus to name a few.

Once a determination of the risk group has been made, the next step is to establish the biosafety level at which experiments will be conducted. The ABSL correlates with the risk group of the agent in most cases, with the lowest level of protection required at ABSL-1 [5]. It is important to note the process for determining the biosafety level at which a particular agent is to be handled. Biological agents are not classified according to biosafety levels; rather, the biosafety level takes into account the containment requirements for a specific agent and experiment based on the risk group category designated at the risk assessment stage. For example, the potential creation of an aerosol for experimental purposes of a RG2 organism may increase the biosafety level from BSL-2 to BSL-2 enhanced or possibly BSL-3. In the biosafety world (and even in IBCs) individuals may get too comfortable classifying agents based on biosafety levels without truly understanding how a risk assessment is conducted. It is therefore important to ensure that the IBC is trained and knowledgeable in these concepts. Subject matter experts on the IBC may already be well-versed in the practice of risk assessment, but it is worthwhile to also communicate to community members the process by which risk group is determined, and how this eventually informs what animal biosafety level is used. Depending on how a given agent is being manipulated, risk group classifications may either be raised or lowered and this in turn can affect their stratification in terms of biosafety levels.

ABSL-1 work involves agents that are well characterized and do not pose any substantive danger to healthy individuals who work with or around these agents. These agents pose a minimal risk to the environment and human or animal health. ABSL-1 facilities are typically segregated from the general laboratory population, with some type of restricted access. This applies to areas within a vivarium or satellite ABSL-1 facilities as well. While most work with agents in ABSL-1 settings may be conducted on the open bench top and generally does not require engineering controls, there are certain circumstances where the latter may be needed. This is one area in which risk assessment is particularly important. The biosafety professional – possibly with the help of the IBC – must do a risk assessment on the agents being used, how they are handled and administered and further take into account what type of animal is being used to prescribe specific engineering controls and appropriate levels of PPE. Engineering controls in this case may relate to requiring administration of the agent within a fume hood or regulating the number of air changes in the room per hour, as stipulated by the *Guide for the Care and Use of Laboratory Animals* (henceforth referred to as *The Guide*), which was developed and is regularly updated by the Association for Assessment and Accreditation of Laboratory Animal Care (AAALAC) [6]. Although many biosafety professionals consider the Centers for Disease Control and Prevention (CDC)'s *Biosafety in Microbiological and Biomedical Laboratories (BMBL)*, currently in its fifth edition, to be our "bible," *The Guide* serves a similar function for those involved in the oversight and design of animal research areas and programs, particularly for animal species not covered by USDA regulations [8].

A wide range of other issues comes into play when conducting work at ABSL-1. Personnel working with animals at this very basic level must receive appropriate training, including identification of the hazards that one may come across when working with the animal species and biological agents being used. A medical surveillance program should be in place even for this basic level of work at ABSL-1, and this becomes all the more necessary when dealing with higher animal biosafety levels. It is critical that areas where animal research is being conducted, starting at ABSL-1, have proper signage on the entry doors to these areas, alerting individuals that they are entering an area where biological research is being conducted with animals. General best practices for microbiological facilities should also be communicated to users of these spaces. There must be areas for disposal of sharps and broken glassware. Laboratory workers should also be trained to avoid recapping needles, as numerous injuries occur during recapping procedures, even when these are done "correctly."

The construction of the animal facility becomes more complex as the animal biosafety level increases. Areas where animals are housed within vivariums should have impervious floors, walls, and ceilings. This is important for animal husbandry, as well as decontamination of an area. These areas should be slip-resistant and impermeable not only to water, but also to chemicals that may be used for either general cleaning or decontamination of surfaces. Sinks and floor drains should be filled with water to prevent sewer gases from entering animal areas, and to prevent the introduction of pests or vermin into these areas [5]. Doors should be self-closing and sealed around their outer edges for similar reasons, and may in some cases have a "sweep" around the base of the door to prevent smaller animals – particularly mice – from exiting the room should they escape their cage while being handled or during animal husbandry. Most animal vivariums should not have external windows to the outside. Windows create security problems as well as the complicating maintenance of appropriate temperature, humidity. and other environmental controls.

Ventilation in animal areas should comply with *The Guide* [6]. Ventilation within these areas must take into account heat and humidity loads, and these systems should not recirculate exhaust air either within the room or to other animal-containing areas. Exhaust air is considered "dirty" and may be contaminated by pathogens or allergens present within the animal areas that could affect both workers and the animals themselves. If it is determined that the potential pathogens present in exhaust air could potentially pose an environmental impact, considerations are made for HEPA filtration. In all animal areas, ranging from ABSL-1 through ABSL-4, it is important to limit the use of horizontal surfaces with regard to both features of the room itself or equipment that is being brought into the area. Such surfaces require much more cleaning, as they tend to accumulate dust and dander, and can act as or facilitate the accumulation of fomites. In this scenario, dust could act as a vehicle for infectious organisms to settle upon; if aerosolized, this fomite could then act as a vehicle for transfer to a new host. PPE can also act as a fomite, which is why gowns, booties, gloves, and other PPE should be removed before leaving a room area, vivarium, or building. It is helpful to conduct a risk assessment and develop SOPs to determine when and where PPE is donned and doffed. It is for this specific reason that laboratory workers are strongly discouraged against bringing their lab coats home, as these may also facilitate transfer of infectious organisms. An animal vivarium should also have a cage-washing system that is capable of reaching a final rinse temperature of 180 F [6]. This applies to either mechanical systems or manual cage washing with chemical disinfecting agents. It is for this reason that most vivariums, especially at larger institutions or facilities, solely use mechanical cage washers for ease of use and efficacy purposes.

We can then build upon the foundational ABSL-1 practices described above to meet the needs of higher animal biosafety levels, where pathogens or infectious substances are in use that are more harmful to the environment, animals, or workers. ABSL-2 is suitable for work involving laboratory animals infected with agents that are associated with human disease and pose moderate hazards both to laboratory personnel and the environment [5]. It also addresses hazards from ingestion as well as percutaneous and mucous membrane exposure. Certain training requirements must be met, and this may involve institution-specific training. At a minimum, training for research personnel should include facility procedures, handling of infected animals and training on the actual administration or manipulation of the pathogenic agents being used in animal studies. Personnel knowledgeable of the potential hazards associated with these agents, as well as the relevant animal manipulation and husbandry procedures, should supervise individuals working at ABSL-2. One of the critical differences between ABSL-1 and ABSL-2 is in the introduction of engineering controls for primary containment at ABSL-2, usually in the form of a biosafety cabinet. These could include Class II A2-type biosafety cabinets that mostly exhaust to the room or Class II B1 or B2 biosafety cabinets that exhaust air to the outside at varying percentages. A biosafety cabinet not only provides a sterile working environment when used correctly, but also offers a primary protective barrier for personnel. When a risk assessment determines that an element of containment is needed, elevation of the agent's risk group and an increase of the biosafety level from ABSL-1 to ABSL-2 are indicated through assessing the risk to humans, animal, environment, or other factors specific to the work being conducted. Other engineering controls may also be administered, but the biosafety cabinet is the most common form of primary containment at ABSL-2. The biosafety cabinet also comes into play when manipulating infectious materials or conducting procedures that have the potential to produce aerosols. Biosafety cabinets may also be used for necropsy, harvesting of tissues, or even animal cage changes if required [5]. If a procedure cannot be conducted in a biosafety cabinet, a combination of PPE and other containment strategies must be used, such as using the actual room as primary containment and possibly considering respiratory protection.

The biosafety professional – in conjunction with the IBC, Institutional Animal Care & Use Committee (IACUC), and institutional veterinarian – can play an important part in researching the shedding rates of biological agents in order to determine when cages or bedding may be considered to contain infectious materials, how bedding or cages should be decontaminated, and when cages or bedding can be deemed to no longer pose an infectious threat. Based on these determinations, it may be appropriate after a certain time period to reduce the animal biosafety level for housing purposes – for example, going from ABSL-2 to ABSL-1. This determination must be made via risk assessment and must also go through the IBC with notification of the IACUC and comparative medicine department.

There is no uniform prescription for ABSL-2 engineering controls or PPE (as is true of any biosafety level), and the purpose of these interventions is to minimize exposure risk. Personnel that conduct work at ABSL-2 should be enrolled in a medical surveillance program [5] that takes into consideration the agents being used, manipulations taking place, and/or animal species involved. Allergies are still an important consideration, even with widespread use of PPE at this level. Animal workers should be advised to report any sort of immune conditions or general personal health conditions that may make them more susceptible to infection. When possible, one should consider specific practices or restraint devices that reduce the risk of exposure for the worker [5].

As with any animal work, animal protocols must be reviewed and approved by the institution's IACUC, as well as the IBC when biological agents are to be used. As discussed previously, it is vital for the IACUC and IBC to work closely and maintain communication with each other. The biosafety professional often serves as a liaison between the two committees, occasionally by also serving as a member of the IACUC. Site-specific manuals must be developed and implemented for animal facilities. The IACUC and IBC should ensure that workers are being trained and made aware of the potential hazards involved with their specific duties. This training needs to be conducted yearly as well as whenever changes are made to policies, procedures, or agents, and must also be documented.

A small but significant difference in signage is required when moving from ABSL-1 to ABSL-2. When working with multiple agents at ABSL-1, it is only "recommended" to post these agents on the door. However, it is *mandatory* to post the names of the agents being used at ABSL-2 [5]. There may be certain agents posted on the signage that restrict an individual from entering due to personal health status or inadequate experience working with the agents, regardless of whether active work is underway. This signage should also include other basic information that is required at every animal biosafety level, such as contact information, PPE entry requirements, and entry and exit procedures.

PPE at ABSL-2 should be specific for the facility. Lab coats, gowns, scrub suits, or uniforms should be worn while working in the vicinity of infectious agents. They should be removed when contaminated, and never leave the premises to prevent unwitting transfer of known or unknown infectious agents. In many cases, regardless of biosafety level, animal facilities have certain practices in place for the donning and doffing of PPE – not only from facility to facility, but even from room to room or hallway to hallway. These facility-specific SOPs may reduce the risk of environmental contamination that could affect other animal populations within these facilities. Gloves are always to be used within animal facilities, but should never be worn or used outside of these areas. They should also never be reused or washed, as this could compromise the integrity of the glove. It is important to provide training on the proper way to remove gloves in order to prevent contamination. Gloves should be disposed of with other potential or known infectious waste within the facility. Hand-washing facilities should be available within immediate proximity of the waste disposal areas where gloves are to be removed [5].

Appropriate waste handling is an important aspect of working at ABSL-2, an oft-neglected aspect of containment work. ABSL-2 requires that a form of decontamination be available for any waste generated from the agent being used, including bedding from caging housing an infected animal. Different forms of decontamination may be used, such as autoclaving or chemical decontamination methods. Careful consideration must be given regarding decontamination of equipment used in experimentation as well as husbandry [5]. Equipment that has the potential to become contaminated must be amenable to decontamination in the event that routine maintenance, repairs or removal from a particular area are required. In the event of spills generated during activities involving infectious agents at ABSL-2, staff or personnel must be appropriately trained to contain and decontaminate the affected areas based on the specific policies or SOPs developed for that facility.

Ventilation requirements in ABSL-2 facilities are more stringent than in ABSL-1 sites. Ventilation in ABSL-2 areas should be calibrated to maintain negative pressure relative to areas with lower biosafety levels. This pertains to animal isolation cubicle areas relative to the outer room, as well as housing or procedure rooms relative to hallways. General exhaust air should always be ducted out of the facility and not recirculated anywhere within the facility [6]. There may be a level of filtration at this level before exhausting takes place, most usually HEPA filtration. Class II Type 2 biosafety cabinets that mostly recirculate HEPA-filtered air back into the room are an exception to this requirement. Biosafety cabinets may also be thimble-connected or hard-ducted through the lab or room exhaust. Biosafety cabinets should be certified at least annually or whenever moved from one location to another [7]. Biosafety cabinets should be positioned in a manner that minimizes disruptions in airflow in the room as well as within the biosafety cabinet itself (e.g., near air vents, doors or high-traffic areas). Finally, ventilation within an animal facility should take into account the humidity and temperature requirements for proper animal husbandry, as established by *The Guide* and other resources.

Unsurprisingly, ABSL-3 procedures further build upon those employed at ABSL-2 since agents used in an ABSL-3 environment can cause serious or potentially lethal disease [5]. In particular, careful consideration must be given to the potential of agent transmission via the aerosol route. Accordingly, ABSL-3 labs require more sophisticated airflow systems, procedures, and PPE. Respirators of some type may be used, depending on the risk assessment. These respirators are typically powered air-purifying respirators or N-95 respirators, which require the user to be enrolled in the institution's respiratory protection program. ABSL-3 workers must also be enrolled in a medical surveillance program within their institution. Beyond the access-restriction implemented in ABSL-2, ABSL-3 typically implements multiple safety and security barriers.

All agent manipulation at ABSL-3 must be done within a biosafety cabinet. Airflow and HVAC systems become very important when making the leap to ABSL-3. ABSL-3 laboratories are required to have an inward directional airflow that acts in a cascading manner: from an outer area into an anteroom, and finally into the lab.

The anteroom is a small room between the actual lab or ABSL-3 area and an outside hallway or common area. In ABSL-3 settings, anterooms are always considered clean areas, and are usually where those entering ABSL-3 laboratories don their PPE.

Animal racks and caging in ABSL-3 areas are usually closed systems in which both supply and exhaust air pass through HEPA filtration. Though there may be positive pressure plenums within these caging systems, measures are put into place to prevent reversal of airflow in these systems, which could release potentially infectious materials into the room. This may be accomplished through an interlock system within the blowers or fans within these animal rack systems. These animal-caging systems should be certified at least annually. Caging ventilation and the redundancies involved in the HVAC system should be tested at least annually to ensure proper function of these systems and prevent reversal of airflow in the event of a mechanical failure.

Laboratory-specific SOPs should be developed for ABSL-3 facilities, including information on the agent, hazard communication, lab-specific procedures, waste disposal procedures, inactivation procedures, a detailed medical surveillance program, sharps use or management, and incident response procedures that at an absolute minimum address spill cleanup protocols. ABSL-3 facilities must have impervious floors, ceilings, and walls. The floors should be monolithic in nature. Impermeable surfaces allow these areas to be sealed and decontaminated if needed. These areas should have no penetrations, but are not required to be gas- and bubble-tight, as would be required in an aBSL-3Ag facility. In an aBSL-3Ag facility this is required due to containment of agents of high consequence to agricultural stocks that could affect an economy. Decontamination within these areas can range from surface decontamination to gaseous decontamination, and the sealing of these surfaces is especially vital in this latter scenario.

Work with large animal species that cannot be housed in traditional containment equipment, such as biosafety cabinets, is conducted at what are known as BSL-3-Ag facilities. In this environment, other modifications are made to aid in primary containment. It is important to carefully control the flow in and out of these labs, in terms of both personnel and airflow, especially when working with agents that may be harmful if released into the environment. BSL-3 Ag facilities implement some of the practices used at ABSL-4 (discussed below), although workers do not use positive-pressure hoses or suits, as they become problematic for workers in proximity to large animals. The key concept is this: in BSL-3 Ag areas, the *room itself* acts as the primary containment, with HEPA filtration integrated into the supply air and exhaust ductwork. Gas-tight isolation dampers can be used in place of HEPA filters for the supply air, but as these dampers may not be reliable, HEPA filtration should also be implemented either in conjunction with supply-air isolation dampers or in parallel, within the supply air as a redundancy measure in case the dampers fail. If a reversal of airflow occurs within the supply air plenums due to the failure of exhausts fans, or failure of both supply and exhaust fans, HEPA filtration ensures that this air will not contain any potential contamination from within the ductwork or elsewhere. Directional airflow, air-pressure-resistant doors with gaskets, air locks, effluent decontamination systems, animal renderers, restraint devices, systems that enable pressure decay testing, and the capability to seal any and all access points in the room are just some of the engineering controls that must be implemented at this level. Pressure decay testing is the administering of pressure to a room or ductwork, and observing the pressure differential to determine the presence of leaks. These areas must provide absolute containment for the protection of both the environment and the public [5]. For all of these reasons, BSL-3Ag facilities are very expensive to design, build, and maintain.

ABSL-4 facilities involve agents that are dangerous and exotic, with no treatment available and high mortality rates for exposed individuals. Differences between ABSL-3 and ABSL-4 include the use of sealed, Class III biosafety cabinets and positive-pressure suits. A complete change of clothes is required upon entry and exit, and personnel are required to shower out upon leaving the lab, which may also be required of ABSL-3+ or ABSL-3Ag labs. Discrete areas within the lab, including the anteroom, should have interlocks.

The risk assessment to determine animal biosafety levels is normally conducted by the biosafety professional in conjunction with the IBC, but it is important to involve the IBC as well. This is useful when there are animal experts or animal model experts on the committee that may have more experience or expertise on a certain subject. The IBC should be responsible for reviewing protocols to make recommendations on containment and animal biosafety levels, and to review work that may pose an unnecessary risk to the personnel involved. IBCs should be composed of individuals with specific expertise, and these experts can provide pertinent information related to experimental protocols that could help determine final recommendations for animal biosafety levels or containment practices.

Transgenic animal models

Transgenic animals that carry a foreign gene – often of human origin – artificially inserted within their genome, have been an integral part of scientific research for decades. Since experimenting on humans is unethical, transgenic animal models can serve as surrogates for the study of human diseases with a genetic component or for infectious diseases for which susceptibility can be modified by certain gene or allele combinations. For instance, insertion of human genes coding for viral receptors may permit infection of the animal model with a pathogen that is otherwise specific to human hosts. Most transgenic animal studies are conducted with mouse models, although smaller numbers of studies are performed with rats and fish. Mice are relatively inexpensive and can be raised in large numbers, their genomes can be easily manipulated, and exhibit many physiological similarities to humans.

It is important for the IBC to note whether transgenic animals are being purchased or are being created at its home institution. The majority of transgenic animals being purchased or transferred fall under the NIH III-F exempt category. This is a category developed by the NIH, which states that any research that falls under the III-F designation is exempt from the NIH guidelines for research involving recombinant or synthetic nucleic acid molecules. Once again there is no singular approach to assigning NIH categories to purchases and transfers; determination is made in conjunction with the biosafety professional and the IBC and should be documented within the IBC meeting minutes.

In the majority of cases, transgenic animals are created at an institution rather than purchased or transferred. Studies involving these animals do *not* fall under the NIH III-F exempt category, but rather into other, non-exempt categories. If the determination can be made that the research, or the purchase or transfer of transgenic animals may be housed at ABSL-1, it is exempt from the NIH guidelines for research involving recombinant or synthetic nucleic acid molecules. The creation of transgenic animals involves use of recombinant or synthetic nucleic acid molecules, which the NIH respectively defines as "molecules that a) are constructed by joining nucleic acid molecules and b) that can replicate in a living cell" and "nucleic acid molecules that are chemically or by other means synthesized or amplified, including those that are chemically or otherwise modified but, can base pair with naturally occurring nucleic acid molecules" [4]. The creation of a transgenic animal typically involves recombinant viral vectors that deliver the transgene that in turn becomes integrated into the genome of the recipient animal. This type of *in vivo* work may not always result in the creation of a new transgenic animal line, but can instead result in a foreign protein being expressed from the transgene. As with studies involving pathogens, the IBC is required to review the administration of recombinant DNA technologies studies. The IBC must also classify them and determine biosafety levels and containment practices, as well as possible dual use research of concern (DURC), gain of function studies, and major actions. Major actions are experiments or manipulations that involve the deliberate transfer of antibiotic resistance to an organism. The new DURC policy has established guidelines for research that falls under categories deemed dual use research and dual use research of concern that will take effect in September 2015. DURC committees can also be either separate from the IBC or a sub-committee thereof, with other subject matter experts potentially being involved.

Generally, the *in vivo* use of recombinant or synthetic nucleic acid molecules falls into one of two NIH categories for recombinant or synthetic nucleic acid molecules; III-E or III-D. NIH III-E classification is for studies that require the notification of the IBC simultaneously with study initiation. Briefly, this includes experiments that do not involve more than two-thirds of the genome of any eukaryotic virus, use a RG1 agent, and are conducted at ABSL-1 [4]. A typical example of this type of work would be using recombinant adeno-associated viruses to express certain genes or proteins within an animal. In contrast, studies classified as NIH III-D require IBC approval before the work may be initiated. This category of studies uses higher-risk agents for experiments involving genomic integration, either *in vitro* or *in vivo*. There are many subcategories of the III-D designation, and the specifics depend on the details of a particular experiment. These subcategories range from III-D-1-a for experiments that involve the introduction of synthetic nucleic acids into RG2 organisms usually conducted at BSL-2 containment, to III-D-7-d which covers antiviral susceptibility and containment. The NIH does a very good job of defining what types of experiments fall into these categories, but it is ultimately up to the institutional IBC to make the determination, with consultation from subject matter experts on the committee and an understanding of the NIH guidelines for research with recombinant or synthetic nucleic acid molecules.

Most *in vivo* experiments will fall under these three categories of NIH III-D, -E or -F, although elements of *in vivo* work may fall under the NIH III-B classification. This designation requires NIH Office of Biological Activities (OBA) and IBC approval before initiation, and encompasses experiments involving cloning of toxin molecules with an LD50 less than 100 ng/kg body weight [4]. NIH category III-A designates experiments that require review by the institutional IBC and NIH Recombinant DNA Advisory Committee (RAC), as well as approval from the NIH Director. These experiments constitute "Major Actions" (as detailed above), entailing transfer of a drug resistance trait to microorganisms that are not known to possess this resistance trait naturally [4]. Clearly, it is important that IBC members know the definitions of and differences between these categories. Part of the role of a biosafety professional is to identify the different types of experiments that may fall into these categories, and to reach out to the IBC to make sure they understand why a particular experiment is being reviewed if it potentially falls into one of these higher-level categories. Institutions that receive NIH funding are required to have an IBC with subject matter experts within the institution as well as community members. It is important to have an active, involved IBC that understands how and why they are assigning these NIH category designations to the work under review.

Regulatory reporting

Institutions that receive NIH funding, have an IBC and conduct research with recombinant or synthetic nucleic acids also have additional important reporting obligations subsequent to an untoward event. Releases of recombinant or synthetic nucleic acid material outside of primary containment, escape from primary containment of an animal that has been administered recombinant or synthetic nucleic acid molecules, or exposure of an individual involving recombinant or synthetic nucleic acid material are all reportable to the NIH. In these cases, the biosafety professional must send a report to the NIH/OBA office upon becoming aware of the incident. They should also meet with the personnel involved in the incident to determine the sequence of events that led to this release, escape, or exposure. Critical details in this process include identifying the particular breach of primary containment, PPE compromise (especially respiratory protection), deviations from SOPs, the extent of decontamination of the area where the incident took place, and any corrective actions that were taken. When dealing with incidents related to transgenic animals or animals infected with recombinant or synthetic nucleic acid molecules, it is important to determine whether PPE was compromised and/or if an animal was able to either bite or scratch the individual. In the latter event, it will be necessary for that individual to receive medical care or treatment at a facility approved by the occupational health program. The personnel involved should let the clinic know the sequence of events, describe any breaches in PPE (including respiratory protection), and be forthright in sharing all relevant information so that the clinic may make a determination of the proper medical treatment, if needed. If the incident involves a breach of primary containment in a

Select Agent laboratory, correspondence with the CDC and a Form 3 (Report of Theft, Loss, or Release of Select Agents and Toxins) must also be submitted in conjunction with the NIH reporting. The reporting narrative should be communicated to both agencies in a timely manner. These agencies will follow up with inquiries regarding breaches in PPE, HVAC function during the incident, recapture of the animal, decontamination of the space, and the current and continued health status of the individual who was exposed or potentially exposed. Incidents that are reported to the NIH are public information and should be communicated to the IBC and documented on the IBC agenda and meeting minutes. The IBC plays an important part in determining potential corrective actions. The IBC can be a useful tool for biosafety professionals, as the different areas of expertise and years of experience of committee members may play a vital role in implementing prevention practices. It is important to have a robust reporting process at institutions that conduct work with synthetic nucleic acids, and open lines of communication and well-established relationships between researchers and biosafety professionals are vital to an effective reporting program.

Select agents and toxins

The CDC defines Select Agents and Toxins as biological agents and toxins that could pose a severe threat to public health and safety [5]. A large majority of Select Agent work is conducted *in vivo* in ABSL-3 or ABSL-4 facilities. Though not part of the original mandate, IBCs are often tasked with Select Agent compliance at research institutions. It is important that specific SOPs be developed for this research, and that all CDC rules and regulations are followed. Many of the practices described above for these animal biosafety levels apply to this type of research. Select Agent research requires sophisticated HVAC systems, restricted access, higher levels of PPE and respiratory protection, medical surveillance, training of personnel in biosafety, biosecurity, and incident response, and having written plans in place. It is also important to maintain animal disposition logs that show when each animal has been infected and when it has been inactivated. Entry into these spaces by approved individuals must be tracked in the form of logs and access records, all of which need to be audited internally. It is important that SOPs for Select Agent labs take into consideration animal biosafety concerns with regard to personnel. These may involve specific PPE requirements, such as puncture-proof gloves if working with sharps or if there are concerns about aggressive animals. On the other hand, over-prescribing PPE may lead to a loss in dexterity or agility, which may result in more harm than benefit. Animal handling, especially with smaller animals, can be a very delicate process where reduced dexterity can compromise the safety of both the animal and the personnel. It is important to take all of these factors into consideration when establishing PPE requirements, and to have a thorough discussion with the lab working with the Select Agent. Many Select Agent experiments may allow use of anesthesia to minimize risks of injury or escape, but not all. For example, some studies conducted with Select Agents may focus on the animal's respiratory or neurological function, such that use of anesthesia may affect the study adversely. In these instances, specific procedures should be developed for animal handling to minimize escape or breach from primary containment in collaboration with the lab, biosafety professionals, and the IBC. The IBC reviews the SOPs for these labs, reviews and approves inactivation protocols, is in charge of Select Agents, and is usually involved with the reporting process if there is a theft, loss, or release of a Select Agent at the institution. The IBC also must be apprised of new labs or Select Agents being implemented at the institution.

Conclusions

IBCs play a vital role in the realm of animal biosafety. The committee is a repository of knowledge, skills, and expertise in a variety of fields that biosafety professionals and institutions can rely on to conduct research with animals that accommodates the animal's well-being, as well as the safety of the personnel involved in the research. Through the risk assessment process, institutions can determine biosafety levels, prescribe PPE, and select appropriate engineering controls to ensure that a successful animal research program can be implemented and executed. It is important that the IBC plays an active role in overseeing all animal work done at an institution, and the implementation of this committee's decisions in work practices will in turn lead to a safe and successful research program.

References

[1] Emes RD, Goodstadt L, Winter EE, Ponting CP. Comparison of the genomes of human and mouse lays the foundation of genome zoology. Hum Mol Genet 2003;12(7):701–9.
[2] Nilsson S, Helou K, Walentinsson A, Szpirer C, Nerman O, Stahl F. Rat-mouse and rat-human comparative maps based on gene homology and high-resolution zoo-FISH. Genomics 2001;74(2):287–98.
[3] The Chimpanzee Sequencing and Analysis Consortium Initial sequence of the chimpanzee genome and comparison with the human genome. Nature 2005;437:69–87.
[4] National Institutes of Health Office of Biotechnology Activities NIH guidelines for research involving recombinant or synthetic nucleic acid molecules. Bethesda: National Institutes of Health, U.S. Department of Health and Human Services; November 2013.
[5] Centers for Disease Control and Prevention National Institutes of Health Biosafety in microbiological and biomedical laboratories, 5th ed. New York: U.S. Department of Health and Human Services; 2009.
[6] Committee for the Update of the Guide for the Care and Use of Laboratory Animals Institute for Laboratory Animal Research Division on Earth and Life Studies Guide for the care and use of laboratory animals, 8th ed. New York: National Academies Press; 2011.
[7] NSF International Standard/American National Standard. Biosafety cabinetry: design, construction, performance, and field certification. NSF/ANSI 49-2008.
[8] Wilson DE, Chosewood LC, editors. Biosafety in Microbiological and Biomedical Laboratories (BMBL), (5th ed.). U.S. Department of Health and Human Services, 2009. HHS Publication No. (CDC) 21-1112. Available from: http://www.cdc.gov/biosafety/publications/bmbl5/.

Crop security: current priorities and perspectives in public sector institutional review

M. Malendia Maccree

Chapter Outline

Introduction

Crop biosecurity encompasses the processes and controls required to protect plant health, human health, and a robust agricultural trade economy. Biosecurity risk assessment for crop health and agricultural production requires case-by-case consideration based on a complex equation of local and global factors. Institutional Biosafety Committees (IBCs) and similar institutional oversight bodies serve an important role in national and international crop biosecurity efforts by ensuring regulatory compliance for crop research conducted by publicly funded institutions. This chapter examines current priorities and factors in risk assessment for crop biosecurity, focusing on the role of institutional review in oversight of public sector crop research

Ensuring National Biosecurity. DOI: http://dx.doi.org/10.1016/B978-0-12-801885-9.00009-3

in the United States (US). Though the specific jurisdictions and processes described are those of the United States, some principles are broadly applicable.

What is crop biosecurity?

Most often, crop biosecurity is defined according to the time, place, and specific stake-holders involved. The term "biosecurity" may be used to refer to basic plant health protection programs as a means to distinguish the processes and goals of plant protection from those of traditional biosafety (i.e., human health protection). The goal of this aspect of crop biosecurity is to mitigate potential for plant disease to occur. Current US crop biosecurity regulations are focused on a small subset of exotic plant pathogens identified as Select Agent pathogens [1], though numerous other endemic and exotic pathogens pose more immediate and actionable risks [2–4]. Crop biosecurity can be extended to include protection against "food terrorism" or threats to the plant-based food supply or commerce in agricultural commodities [5]. Fresh produce remains an on-going source of foodborne illness, leading to increased funding of plant-based food safety research at universities [6]. Institutional security programs aimed at preventing intentional vandalism or theft of experimental crops (and of associated intellectual property) may be included under the umbrella of crop biosecurity [7]. As outdoor testing of genetically modified (GM) plants and international importation restrictions on genetically modified organisms (GMOs) become more commonplace, biosecurity programs have expanded to focus on preventing and detecting adventitious gene flow and seed mixing [8]. Regardless of the specific threats and vulnerabilities examined or mitigated, the common larger objectives of crop biosecurity programs are to avoid negative impacts in domestic and export commodity markets and assure smooth functioning of the global agricultural crop economy. For the purposes of this discussion, crop biosecurity is defined in this broadest sense to capture the innovative landscape of public sector research under the purview of local biosafety committees. Other terms are used as defined in the ISPM glossary of phytosanitary terms [9].

Prevention, protection, and accountability

Crop biosecurity programs, in general, are focused on three functional goals: disease prevention in plants, human health protection, and regulatory compliance for GMOs. Each goal presents unique challenges and deserves independent examination to identify how and where local oversight bodies serve a role in ensuring crop biosecurity.

Disease prevention

Plant disease prevention

Disease pressures on any particular crop are multilayered and varied in impact from location to location. These pressures consist of exotic and endemic biological factors (pests and disease) as well as local abiotic factors (e.g. weather, cultural conditions, and chemical agents). While biosecurity programs are focused in biological disease agents, abiotic factors also play a role as environmental influences in disease and are discussed later in this chapter as a component of risk assessments.

Research involving endemic disease agents and plant-associated microbes can present a challenge for oversight. No regulations or official controls may be imposed for locally endemic pests. Nonetheless, it may be imperative to control local pests when employed for experimental use and to prevent inadvertent transport of regulated non-quarantine pests (RNQP) from their area of origin. Exclusion (restricting an organism through quarantine or removal) can become meaningless for locally abundant pests that are regarded as a tolerable or inconsequential nuisance in local crops. Programs for control of endemic pests often focus on integrated management of pests rather than complete elimination or exclusion. Scarabaeid beetle species are an example of an important group of insects that cause damage to crops worldwide as exotic and endemic plant pests [10]. When endemic to an area, they can be absent for years and then occur in outbreaks that are hard to control. Areas of intensive crop monoculture can serve as incubators for pesticide-resistant pests. Also, overuse or poor choice of chemical controls can produce resistance in larger, rapidly evolving pest populations, making locally abundant populations hard to control. If endemic disease cycles are ongoing and predictable, routine monitoring of indicator crops is possible and control measures (such as pesticide application or crop removal) can be targeted accordingly. In cases where pesticides and exclusion fail to control endemic pests, species-specific biocontrol measures can show promising results as seen in the case of Scarabaeid beetles. It is important to consider that the known endemic disease agents may represent only a subset of the actual disease agents present at a given location (i.e., additional species may be present, yet unidentified). Introduction of new hosts can lead to discovery and unintentional magnification of previously unknown latent disease reservoirs. Economically important crop threats, such as Pierce's disease in California grapevine, highlight the effect of introduction of highly efficient insect vectors (Glassy-winged sharpshooter) and increased population of susceptible hosts into an existing latent plant pathogen reservoir [11]. The key challenges of working safely with endemic pests are to maintain appropriate vigilance in controlling import or export of pests, and manage existing populations to prevent worsening of conditions or decline of effective controls.

Preventing introduction of exotic plant disease agents into new locations is the primary goal of national and international plant health regulations and quarantine efforts. The United States, similar to most developed and developing nations, has a robust system of state and federal authorities to govern the importation of materials into the US and movement within the US from state to state [12]. The US may be unique in the degree of regulation and oversight in movement of agricultural materials. Federal permitting is required for interstate movement of materials under regulations enforceable by USDA. If specific local restrictions do not exist, control may be elevated to the federal level through a state-located plant health director (SPHD). Federal and state agricultural officials can respond to inquiries regarding transit of agricultural materials through a given area and may issue specific activity or use permits pertaining to movement of items under quarantine. Permitting and notification programs serve as means to maintain lists of registered research entities, and the permitting process usually includes facility assessments or inspections. The permitting and notification processes provide an effective means of monitoring the known and established movement of high-consequence exotic pathogens and pests in the US. Permitting programs are especially important for plant pathogens because no other specific requirements or guidance may apply under existing commerce or transportation (e.g., DOT, DOC, or IATA) regulations if an organism is not subject to export controls, genetically modified, or known to cause disease in humans or animals. Failure to follow established regulations and gain the required authorizations can result in monetary fines as well as shipments of sensitive research materials being delayed, unintentionally released (if opened by unauthorized parties), or destroyed in the shipping process. Safe use of exotic pests in research is often dictated by prescriptive written or regulatory controls; interpreting and adhering to strict regulatory controls can be complicated.

Historically, plant protection efforts have focused on monitoring and controlling routes of trade and movement of commercial material as the primary pathways for introduction of pests. However, with increased global mobility in our society, the available pathways for pest introduction have expanded and the greatest risks are posed by deliberate importations via personal mail or passenger baggage [13]. Crop research scientists are global travelers and have historically relied on open sharing of microbiological agents, plant seeds, and curious souvenirs in the course of research. Requirements for monitoring and regulation of transfer of microbial agents are relatively new to microbiological research and may not have become the norm within crop research. Recent highly publicized mistakes involving storage and shipment of infectious human pathogens [6] and embarrassing retractions of research publications in plant pathology laboratories [14–16], also point to the potential for unreported errors and research accidents involving transport and use of lower-hazard plant pests. Local oversight entities, such as IBCs, are charged to monitor laboratory activities and support researchers in compliance with regulations. Beyond research labs, the response to a 2012 disease outbreak in Los Angeles, California, illustrates the impor tance of public outreach [17,18]. Diseased plants brought to the US by suburban fruit tree enthusiasts threatened the state citrus crop with extinction. Education and outreach efforts aimed at nursery industry audiences failed to anticipate and address an important threat posed by home gardeners. This unexpected event necessitated an emergency *ad hoc* program to educate and solicit cooperation from the general public and nonindustry audiences [19]. Local oversight bodies are ideally placed to pay attention to the growing need for outreach activities aimed at researchers and general public audiences. Biosafety committees operating within academic institutions are particularly well-poised for meaningful and effective action.

Plant disease response

Crop disease monitoring and response programs are crucial for protecting agricultural resources at all levels. These functions are performed by cooperative agreements among federal, state, and local agricultural authorities throughout the US. The key to an effective and rapid response to pest introductions or disease is a well-coordinated resource network. The National Plant Diagnostic Network was founded to coordinate and advocate for such efforts in the US [20]. Diagnostic, quarantine, and research facilities linked by a robust web of information technology infrastructure are integral to a plant disease response network. One physical facility may serve more than one of these functions or they may be separated into distinct entities as needed for the complexity and impact of plant disease in the local area. Resources committed to such efforts can vary widely by location and priorities in plant protection.

Diagnostic facilities serve the vital functions of detection and verification of plant disease; these facilities may rarely house a pathogen of great concern, but must maintain vigilance (and compliance) for all possible anticipated pathogens and assure a proper chain of custody for sensitive samples. The primary function is always to accurately identify or classify the organism in question. The specific manner of detection used in the lab is a factor in determining risk posed by the biological agents. Modern nucleic acid testing can be performed on nonviable plant rubs and filters prepared in the field; more complex bioassays or graft tests require establishment of disease as part of the diagnostic method [21]. Disease outbreaks are more likely to involve movement and handling of diseased materials at volumes greater than what routine monitoring programs are designed to accommodate, so increased capacity for storage and shipping must be planned in or available upon demand. Holding and processing large amounts of infectious field samples can present a contamination threat to sensitive diagnostic tests; therefore it is often common for diagnostic functions to be separated from quarantine or research functions. Quarantine facilities are usually maintained for safe holding of larger amounts of infected or infested materials than a diagnostic facility. Quarantine facilities often support diagnostic and research facilities during incident response as triage and staging location for receipt of infested materials or holding capacity for diagnostic samples of concern. Quarantine facilities are designed to contain the worst high-consequence pathogens and pests conceivable in outbreak-proportion volumes or concentrations. The primary function of a quarantine facility is containment of biological organisms: dead or alive. However, such facilities may be *ad hoc* space acquisitions in a time of crisis and might not be designed for ongoing research investigations. Diagnostic and quarantine facilities may be operated under specific blanket function-based or incident-based agreements with agricultural authorities, and often rely on allocated governmental funding streams for maintenance of equipment and staff salaries in order to serve critical plant protection functions.

Research facilities differ from diagnostic and quarantine facilities in that they are designed to conduct high-hazard pathogen or pest research in controlled conditions. Research facilities function to support exploration of scientific questions and testing of hypotheses regarding plant disease. Activities in research labs can vary widely and projects often are dependent upon unpredictable grant funding streams. Successful plant and plant pest growth are the priorities for a plant pathogen research lab. Research laboratories may include indoor and outdoor plant growth areas, where infected or infested materials are safely stored and allowed to grow and thrive. Plant pathogen research laboratories vary widely according to the pathogens studied and the plant models in use [22]. Research laboratories are commonly responsible for initial investigation and development of diagnostic tests and tools for exotic organisms, which can then be used for plant disease monitoring and detection in the field [3]. Often research conditions will mimic those that are ideal for disease to support optimal infection of healthy plants, but conditions are highly controlled and volumes and concentrations of pathogens or pests in use may be higher than found in nature (though not necessarily more hazardous). Testing of Identity Preserved [23] or GMO plant lines for disease resistance is a common activity of research labs that requires experimental plant infections with disease agents. In such cases, the plant lines (in addition to regulated pests) may have requirements for containment or segregation from local plants. While diagnostic and quarantine labs may operate under state authorities, high-hazard plant research facilities are more often under the purview of local oversight committees such as IBCs.

Human pathogens on plants

Fresh produce accounts for more than half of all foodborne illness in the US according to recent estimates reported by the Centers for Disease Control and Prevention (CDC) [24]. While the examples of specific attacks are few, there are enough data on the dire health consequences of crop contamination to indicate that a disease detection and response network is essential to protecting human health regardless of the source of contamination [5]. Disease outbreaks have increased along with the growing demand for fresh produce and intensification of food production. These human disease outbreaks are still studied on a case-by-case basis with each outbreak exposing new vulnerabilities and lessons learned. Similar to plant disease outbreaks, data from case studies yield valuable clues for prevention or detection of human disease outbreaks. Public sector research has played a major role in the development of new food safety regulations and continues to be integral to understanding disease outbreaks [6]. This vital real-world observational knowledge informs and adds upon experimental research to contribute to the development of evidence-based best practices for industry.

Whether the need is outreach and education, or support of detection and response measures, local oversight entities such as IBCs contribute specialized expertise and facilitate institutional aims in the battle to control or eradicate diseases caused by biological agents.

Biotechnology research oversight

Crop development and plant-based research

Assurance of a sustainable and nutritious food supply for a growing global population requires crops with improved ability to resist pests and disease. Thus, current priorities in crop development and plant-based biotechnology are a factor to consider in crop biosecurity [25]. New technologies applied to crops require extensive evaluation to assure that the crops themselves and any crop-derived products are safe for unconfined, outdoor use. For many agriculturally and economically important crop plants, development of disease-resistant plant lines is the only enduring and environmentally sound means to control and prevent disease. As disease agents evolve, climates shift, and food insecurity increases globally, GM crops may be the only hope for introducing new disease resistance and hardiness genes into some crop species. In parallel with efforts to combat plant disease with durable host resistance, GM plants may be useful in addressing important medical, veterinary, and environmental health issues. Plant-based technologies for production of vaccines in plants [26] and use of GM plants to address environmental contamination [27] show promising results in providing safe and environmentally beneficial solutions to support global health. Plant biotechnology shows promise for addressing global health concerns [28] and is recognized as the most rapidly adopted technology since the beginning of agriculture.

Despite decades of research and successful safe application of GM plant technologies, an evolving and somewhat discontinuous regulatory landscape, as well as public distrust, continue to pose significant barriers to adoption of many new technologies. Guidelines published by the Codex Alimentarius Commission in 2003 [29–31], provide basic worldwide ground rules for food purity and sanitation, which apply to GM crops. However, varied approaches in review and authorization of GMOs among different countries have created a contradictory and fragmented array of guidelines and regulations. Many nations worldwide united under the principles of the Cartagena Protocol [32] in 2000 as the source of guidance in the use and commercialization of living modified organisms (LMOs). "The Protocol seeks to protect biological diversity from the potential risks posed by living modified organisms resulting from modern biotechnology." A key point of protocol is the right of the importing nation to follow a precautionary approach in prohibiting importation of living-modified organisms. "Lack of scientific certainty due to insufficient relevant scientific information and knowledge regarding the extent of the potential adverse effects of an LMO on biodiversity, taking into account risks to human health, shall not prevent a Party of import from taking a decision, as appropriate, with regard to the import of the LMO in question, in order to avoid or minimize such potential adverse effects." North American nations, such as the United States and Canada, adopt a strict evidence-based approach requiring scientific data to confirm any plant, human, and environmental health risks posed by the GM crop. Differences in how countries evaluate risks posed by LMOs/GMOs result in differing criteria for what is allowable for importation and exportation [33], with some nations instituting zero tolerance for any GMO and others allowing some level of contamination as acceptable. Asynchronous approval of GM plant products in various markets coupled with trade bans on GM crop products have resulted in disruptions in commodity prices and trade [34,35].

The US regulatory framework is parsed among four federal agencies according to research venue and impact of GMO product [36]. The National Institutes of Health Office of Biotechnology Activities (NIH-OBA) issues guidance for use of GM plants in labs and greenhouses and requires review of research involving GMOs at NIH-funded institutions [37]. The Department of Agriculture, Animal Plant Health Inspection Service (USDA-APHIS) is the lead agency to set and enforce regulations for outdoor release, interstate transfer, and importation of GM plants [38]. GM plants to be commercialized as food or used to produce drugs are reviewed in a voluntary process with the Food and Drug Administration [39]. Outdoor use of GM plants for bioremediation or production of industrial or insecticidal compounds is reviewed under the Environment Protection Agency criteria [40]. As crop technologies grow and evolve, increasing gaps and inconsistencies in US and international regulations have confounded research progress and commodity trade [34]. Institutional Biosafety Committees play a key role in risk assessment during early development of novel biotechnology tools as well as routine research applications of established technologies. Beyond technical safety review of research, public transparency mandated by NIH for institutional review provides an important avenue for community stakeholder involvement in biotechnology research.

Risk assessment for GMOs

Most publicly funded universities and research institutes in the US adhere to *NIH Guidelines* for biosafety review and containment of GM plants and plant pests. These guidelines are largely based on plant health considerations for use of GM microbial agents with whole plants in laboratory and greenhouse environments. In order to best elucidate the risks and regulatory requirements for plants and plant pests that have been altered through biotechnology techniques, it is helpful to examine plants and plant pests independently, factoring each together with environmental conditions to form an overall risk profile for the work. It is important to note that the total plant disease risk calculation, factoring plant host, microbial agents, and environment, is multiplicative, not additive [41]. The presence of essential factors, such as a receptive host, is required for a risk probability to be truly relevant. Unlike risk assessment for human or zoonotic biohazards, the host (plant) may be absent from the experiment or the local environment may be entirely unable to support host pollination or infection. If one factor is effectively eliminated, a zero value enters the equation and the risk of disease spread or gene flow may be mitigated nearly to zero. The immediate plant disease or gene flow risks posed by most public sector plant science research are vanishingly remote. This low-risk profile for most research projects is often due to the absence of susceptible or receptive plant hosts or limiting work to times or locations where environmental conditions that do not support pathogen infection or pollen transfer. However, the absence of immediate risk does not indicate that experimental research is entirely free of risk. In addition to the consideration for the immediate local environment, the risk assessment must consider the broader impacts of an accidental release, including possible consequences of research materials entering the local environment, agricultural commerce, or the food chain.

Similar to human and zoonotic pathogens, modifications made to plant pathogenic agents are always a major consideration in risk assessment for plant-based research. Modifications to plant pathogens should always be reviewed for potential to increased adaptability to new hosts or environments. Impacts may be inconsequential, quite severe, or unknown. Changes in host range must be reviewed for potential impacts upon new or unknown hosts. Highly infectious plant viruses that do not illicit lethal or detectable disease in plant hosts have proven extremely useful as heterologous expression tools and abundantly shared in academic research communities [42]. Though plant viruses have not been documented to adapt to and cause disease in humans and impacts in plants may be inconsequential, the infectivity of recombinant viral vectors should be considered in risk assessment and containment plans. Fungal pathogens present highest risks for adaption to colonize animal hosts and for persistence in the environment because many are free-living and safe eradication options are limited. Bacteria, to a lesser extent than fungi, may adapt to infect animal hosts or immune-compromised workers. In addition to changes in host range, alteration of plant-associated microorganisms used in whole plants must consider the likely pressures and avenues for exchange of genetic information among microbes within the plant host or resident soils. Are plants grown from clean soil and seeds? If exchange of genetic elements is possible among closely related species, what changes could occur to pathogens in the course of plant infection experiments where more than one pathogen may be present? If new or exotic pathogens could be formed, what are the most likely causes and consequences of an accidental release?

Apart from considering the microbial agents that may be in use on a plant, the plant itself may be a factor in the risk assessment. Noxious weeds require careful review whether genetically modified or not. These excessively weedy or invasive plants are a major environmental concern in many parts of the US, particularly in local waterways and wildlife habitats. Any attempts to modify these plant species may result in increasing their negative properties. Plants such as potato and cassava create natural toxins that are hazards for human consumption. These toxic compounds are evaluated whenever plant species are modified through conventional techniques or biotechnology. One plant species of note, *Ricinus communus*, naturally produces a highly regulated and dangerous toxin, ricin. Any modifications to this plant or other similar plants that produce toxic or irritant compounds should be considered in a risk assessment. Any proposed research involving Select Agents and Toxins, such as ricin, is subject to specific federal oversight and approval.

NIH guidelines do not provide great detail on GM plant (i.e., pollen and seed) containment strategies. This is in keeping with the scope of laboratory and greenhouse research, where properly contained experimental plants are extremely unlikely to impact nearby outdoor crops. Nonetheless, modifications made to plants require critical and individual attention for review at this early stage. Routine studies involving GM plants expressing nontoxic plant genes present little concern for environmental or plant disease impact if the transformation agent has been eradicated from the plant, and the plant is maintained in a contained environment separate from other plant lines. However, the NIH guidelines do specify a small number of instances in which plant research must be approved by IBC or NIH before initiation of research; these are: experiments involving expression of genes from human pathogens, production of toxic compounds (LD50< 100 ng/kg), and production of infectious agents (animal or plant pathogens) in plant hosts [37]. In general, expression of nonplant genes in plant systems raises the level of concern for impacts to the environment if plants are accidentally released. Plant pest sequences expressed in plants must be reviewed for potential to create new plant pests or complement function for impaired pests. For example, expression of RNAs from viral plant pathogens in plants can result in creation of novel plant pathogens. Industrial or pharmaceutical compounds produced in plants may present little to no risk for plant health or human health, but result in negative economic impacts for local farmers in the event of comingling or accidental release of seed in the course of research operations.

The third factor to consider in risk assessment is the environment in which the plant will be grown and stored. This aspect of the risk assessment may be the most valuable contribution of local review. The guidelines for review of research involving plant pests and GMOs require a subjective evaluation of detrimental effects to local crops. It is within the purview of each local oversight body to apply this standard to review of research activities. Determining which type of research activities constitute a threat to agriculture depends upon the local crops and operations. Gene flow is only possible if sexually compatible species are within range. Most pathogens require water, insect, or human intervention to find new plant hosts. There are many easily exploited environmental barriers to gene flow and spread of plant disease. Thus, an open-ended performance-based (nonprescriptive) approach affords maximum flexibility to avoid unduly encumbering research. However, unless USDA has issued specific requirements under permit, local reviewers may find themselves in a quandary as to how stringent controls must be in a given locale or research application.

Risk assessment for research involving destructive plant pests and GM plants is one of the most complicated and challenging tasks the IBC will face. The requirement for considering each factor independently and then all together in the experiment, necessitates a fair amount of technical expertise applied in a case-by-case and time-intensive method of review. Accessing and maintaining this expertise within a local oversight body is a challenge that IBCs may struggle to address.

Risks, challenges, and opportunities in institutional oversight

It is important to consider the difference in perspective and full impact of consequences that apply to risk assessment for local institutional oversight entities. While national programs in crop biosecurity aim to secure the nation's crop food supply at a high level, the primary mission at a public research institution is to avert individual incidents of unauthorized release beyond the bounds of established containment or confinement areas. Of equal importance, the IBC mission of compliance must be aligned with the other missions of public sector science: research, education, and outreach.

Risks and consequences

The primary risk event to avoid is unauthorized release of plant pests or regulated GMOs. The risk event may be the result of human error, intentional acts, natural disaster, or mechanical failure. The ideal institutional oversight program must include effective means to detect, prevent, and respond to an unauthorized release. In developing a risk management plan, it is important to examine the possible consequences and assure that input from all stakeholders is considered. Unless locally relevant drivers are behind crop biosecurity programs, prioritizing compliance with abstract or ill-conceived regulations amid research directives can be challenging. Examining the full breadth of risk posed to the institution and using effective means to convey the identified risks to key decisionmakers is essential to soliciting and maintaining support for a robust institutional oversight program.

Consequences of unauthorized release can be examined in five dimensions: health and safety, operational, reputational, financial, and strategic. While the health and safety risks of GM plants are hotly debated and often poorly understood by the nonscientific public, decades of research support the assertion that GM crops pose no greater risk to human health than the production of crops through conventional methods. Institutional review of the details regarding manipulations of gene and genetic constructs, as required under NIH guidelines, is often the first level of review that most publicly funded researchers will face in development of an improved crop. Each institution has authority to restrict or deny unsafe and hazardous projects involving GMOs. Operational consequences of a release are damage to the institution's ability to carry out the core mission. These could include restriction or revocation of funding or the ability to conduct research with regulated materials. If reputational consequences are considered, then the financial impact of losing grant competitiveness or failing to attract the best students and faculty must be factored into the equation. It follows that financial consequences of an incident involving GM crops can be unlimited in size and scope, impacting a public institution at all levels for many years. Initially, there are costs of fines and mitigation to consider. However, it is impossible to predict the downstream financial consequences of reputational and operational consequences. Local crop losses may occur due to market factors or under USDA mitigation (crop destruct) orders. Beyond the overtly negative consequences of an unauthorized release, it is essential to also factor the strategic risks and losses involved. Strategic risks are defined as the cost of opportunities and collaborations, which are lost as a result of an incident. Often, strategic risks are left out of the equation in considering the true cost and benefits of ensuring compliance oversight at an institutional level. Institutions that do not rise to the level of control and oversight required for high-level plant pest and GMO research may lose competitiveness for important funding streams. The strategic risk of neglecting opportunities for public outreach is also discussed later in this chapter.

Challenges in local oversight and review

Navigating regulatory frameworks

The framework within which an IBC operates depends upon how it is defined. Depending upon the breadth and type of plant-based research that is conducted and supported by the institution, the jurisdiction and scope of local oversight can be difficult to determine. The IBC may serve strictly as defined by the NIH guidelines to review research involving only GM organisms in contained environments such as laboratories and greenhouses. However, in practice, it is often hard to extricate outdoor research and use of all manner of plant disease agents from the risk assessment for use of GM plants and other GMOs in applied agricultural research or basic plant science investigations. Human pathogen research conducted on plants and in plant growth and food processing settings also presents unique dilemmas to the standard institutional biosafety review. Institutional authorities are increasingly expanding traditional biohazard review to plant growth areas and supporting compliance oversight activities for quarantined and regulated plant disease agents that are not genetically modified. Environmental (outdoor) release of GMOs and other biological organisms may also require institutional oversight to assure regulatory compliance and incident response requirements are fulfilled. Authentic and effective review of such research activities and regulated materials presents a conundrum to an IBC charged to operate under the NIH guidelines which affords little guidance for coordination of combined review and oversight mechanisms.

Though each of respective federal and state agencies may issue guidance and regulations pertaining to the articles under the purview of the agency, it is the task of the researcher and institutional reviewers to assess how and where the individual mandates apply to the research. With proliferating compliance requirements governing shipment, transfer, and handling of plant materials, the many competing and overlapping requirements become difficult to reconcile. Funding avenues and availability have also influenced the form of public sector crop research. Competition is fierce for limited research funding in the field of applied crop-based research. Compliance burdens are increasingly recognized as the key hurdle preventing more public sector science from being translated into agricultural solutions [43]. Impacts of gaps in operations or a stalled review at one institution may affect projects at multiple institutions and relationships with industry collaborators. Beyond procedural obstacles and delays, failure to comply with USDA regulations can result in monetary fines or restriction of future research. While institutional biosafety committees are charged to review research under the NIH guidelines, adherence to all other federal regulations often becomes a greater challenge in the case of plant-based research.

Outdoor research involving GMOs and experimentally inoculated crops may fall under the purview of local IBCs and institutional oversight committees, or may be deferred to field operations groups. Land grant universities and agricultural extension stations are employed to host all manner of outdoor research, yet NIH *Guidelines* are limited to the description of indoor containment. Any movement of GM plants or seeds outdoors raises the risk of accidental or intentional release or commingling with commodities. Vulnerabilities in intrainstitutional operations may present avenues for accidental release or intentional sabotage of research by activists. Research collaborators may or may not be onsite for evaluations. Specific site information and review is required by USDA-APHIS prior to initiating any outdoor research involving GMOs in the United States. It is also important to consider that outdoor field research is rarely limited to outdoor research locations only. Field-harvested materials must be collected and stored in a safe contained location. USDA-regulated articles held in greenhouses and laboratories may fall under the purview of the NIH *Guidelines* for IBC review. However, these materials remain subject to USDA regulations that require a higher level of segregation and identity preservation than specified by NIH.

As academic institutions are host to pioneering work in plant pathology and crop development, IBCs must often interpret and apply ethical, operational, and safety requirements in areas of research where industry guidance is lacking and clear regulations have yet to be established. New technologies in plant genome engineering are beginning to blur the distinctions used to define GM organisms [44]. Nuclease-based genome-editing systems provide precision in the location and type of genome modifications and now allow for modifications to plants that would be essentially indistinguishable from those which occur naturally [45]. Plants produced with these technologies may be considered unregulated in some countries, while others consider them GMOs. Also in development are crops that utilize GM rootstocks or RNA interference technologies to avoid introducing any foreign genetic elements to the edible portions of plants. These emerging technologies are widely studied and developed in public sector research prior to implementation in commercial crops. Research involving human pathogens on crops is another area of uncertain regulatory coverage. Guidelines for the use of human pathogens in laboratories do not serve well in greenhouses and fields. Still, risks involved with handling plant materials and soils that contain human biohazards and restrictions on feasibility of experimental plant inoculation models must be considered. General health code requirements and regulations prohibiting illicit discharge of biohazards may apply, but no specific regulations or guidelines exist to govern confined outdoor release of human pathogens for experimental purposes.

In responding to the complex requirements of modern research, IBCs may serve as a hub for consolidation of regulatory compliance oversight and serve the role of facilitators of research compliance. Obstacles in interpretation and application of regulations and guidelines are a growing concern as research and compliance systems grow in complexity.

Infrastructure management

Local institutional authorities often oversee the management and operations of diagnostic, quarantine, and research facilities. Plant protection programs are largely publicly funded in the US, with university facilities and subject matter expertise being essential to the enterprise. Operational requirements for high-containment plant pest facilities tend to be highly specialized based upon intended use. Published guidelines exist, but ultimately local pest and plant host circumstances are the best guide for effective and necessary controls. Understanding and prioritizing plant health and crop security needs within IBC programs designed for human biohazard control can be challenging without input from plant protection specialists. Many of the administrative processes and sampling mechanisms used for response to plant disease outbreaks also serve well for response and detection of human pathogens on plants. However, standard plant pathology labs may lack the biosafety controls required for investigations involving human pathogens, particularly those which are more at risk for exposure if aerosolized (e.g., BSL2 equipment and practices). Local oversight committees can serve as the definitive and locally accessible venue to reconcile shared use requirements and best practices for facilities by uniting subject matter experts with institutional authorities.

While threat identification and vigilance are important in maintaining security of research operations, distinguishing between intentional and unintentional incidents is of limited utility in the case of plant health and crop protection concerns. If either type of action goes undetected for long, it will be more difficult to find the culprit or remedy the cause. Prevention of pest and pathogen introductions to crops or mishandling of regulated seeds and plants often starts with a robust detection and response network. This extends to institutional oversight of operations. The ability to detect and respond to research mishaps within an institution was raised to prominence in recent public disclosures regarding misidentified or mishandled biological materials at public research institutes [46]. Compliance programs that function following process-oriented review, which have long been the standard of seed industry and agricultural operations, are easily suited to institutional biosecurity review. In taking this approach, one looks at each step in a handling process to identify where accidents (such as accidental crop contamination or pathogen release) are most likely to occur. Then those vulnerabilities in the process are targeted for monitoring and corrective actions if needed. Documenting the whole process is also essential for monitoring its overall effectiveness and identifying gaps in control systems. For example, failure to communicate complete and accurate information is often a gap identified in process analyses and incident investigations. Process-oriented oversight allows for identification of operational vulnerabilities and threats, facilitating more efficient targeting of institutional resources. The ability to describe the various control systems at work within an institution will also enable better application of a diverse array of regulations.

Public outreach

To assure authentic review of plant-based research, IBCs depend upon project-specific risk assessments reviewed by subject matter experts familiar with plant and plant pest containment. Plant science expertise is explicitly required on the IBC under the NIH guidelines if plant-based research is conducted with GM plants or plant pests. Recruiting and retaining volunteers to serve this capacity for the institution can be a struggle for some entities. Increased opportunities for outreach and alliances with members of plant pathology societies, plant protection agencies, and crop commodity groups would aid institutional officials in gaining access to a larger pool of volunteers with relevant expertise and motivations. When recombinant or synthetic DNA research includes outdoor use of GMOs, knowledge and expertise in field research may also be required on the IBC. While not all stakeholders are suitable or available for participation in institutional review of research, developing suitable means for soliciting advice and input from local farmers and commodity interest groups when needed is essential to making locally relevant determinations on research.

The richest opportunities shared among US public sector research institutions are born of the essential goals to support research progress and facilitate public education [47]. It is important to note that while ethics is recognized as an important aspect of review for animal and human subject research, the ethical landscape of plant research is not precisely defined. What uses of this research are supported by our society and how can the general public best support the progress of publicly funded research towards achievements which are useful to humanity? Raising the questions of ethics and societal benefit in review of public sector crop research may be an important step in resolving the current crisis in public trust and fear around adoption of plant-based biotechnology. As the public debate rages over GM plants, it is time to reconsider the vital role that transparency and stakeholder involvement have in GMO research, and how IBCs serve the public trust. Much attention has been directed towards the "what" of GM plants without the public being given the opportunity and invitation to consider the "why" of plant research. Why is it vital to plant health that we insert specific genes into plant genomes? What benefits to agriculture or humanity can GMOs provide? If more IBCs had active public education campaigns, would citizen and press requests for public records become irrelevant? Understandably, many IBCs approach the recruitment of community members with great caution, due to sensitive and confidential information that may be shared during review of projects. However, a more proactive and projective approach to transparency can be achieved with educational outreach and recruitment of community members to an IBC. Educational institutions in particular are ideally poised to facilitate public involvement in the review process; local review can be designed to include public input as a participatory opportunity in science education, which also supports the academic mission of research, teaching, and outreach.

Risk management opportunities

Local research oversight bodies navigate a harrowing path between the struggles of resource limitations and growing regulatory burdens to facilitate safe and responsible research. Including the considerations of a diverse group of stakeholders in mitigating risks of potentially global impact presents a challenge. A more complete and extended accounting of potential risk and benefits is often necessary to the risk analysis. For example, the reputational and operational threats posed to the institutional mission must be factored into the cost of recovery from a potential incident. Negative press and loss of reputation impact recruitment of new research faculty. Quarantine measures or crop destruct orders can have dire effects on research activities and agricultural commodity production. The risk of failing to take proactive approaches in public education and outreach can include an increased likelihood of public distrust, protest, and vandalism directed at GM crop research. The diverse array and interests of the stakeholders also necessitate a different approach to management of the risks, which are identified. Enterprise risk management approaches that take into account uninsurable and intangible risks, such as missed funding and faculty opportunities, and also reach throughout the entire enterprise to look for strategic ways to address risks, may represent the most sustainable and agile models [48]. While universities with decentralized hierarchies can face the greatest struggles in effective research oversight, it should also be noted that public sector research and university settings especially favor an enterprise risk management approach due to the availability of vast resources [49]. The average land grant university and associated public sector research systems embody a vast base of expertise and knowledge in all aspects of plant health, crop production, security, and systems management. Taking an active and engaged enterprise risk management approach in research compliance is likely to identify the right stakeholders, bring them into the effort, and accommodate the needs and involvement of all stakeholders on the appropriate level.

Conclusion

Effective biosecurity is an on-going process, not a product or a static result to claim accomplishment by completion. Institutional review serves the critical role of coordinating and applying the relevant local knowledge, ethics, and workforce to the process of biosecurity. While progress is needed in public transparency and complicated regulatory landscapes continue to challenge oversight and research objectives, significant outreach and education opportunities also exist in institutional review. With funding for public education and public sector science research continually challenged, the question of ensuring crop biosecurity for many public institutions becomes lost in larger struggles of financial sustainability. An inclusive enterprise-wide risk management approach to leverage all available resources is a promising option supporting crop biosecurity objectives within public sector research.

References

[1] United States Department of Agriculture (USDA) Animal Plant Health Inspection Service (APHIS). Posession, use, and transfer of select agents and toxins. 7CFR part 331. May 12, 2014; 2014.

[2] Schaad NW, Shaw JJ, Vidaver A, Leach J, Erlick BJ. Crop biosecurity. APSnet features. Online. St. Paul, MN, USA: American Phytopathological Society (APS); 1999.

[3] Barkley P, Shubert T, Schutte GC, Godfrey K, Hattingh V, Telford G, et al. Invasive pathogens in plant biosecurity. Case study: citrus biosecurity Gordh G, McKirdy S, editors. The handbook of plant biosecurity. Netherlands: Springer; 2014. p. 547–92.

[4] Magarey RD, Colunga-Garcia M, Fieselmann DA. Plant biosecurity in the united states: roles, responsibilities, and information needs. Bioscience 2009;49:875.

[5] Khan AS, Swerdlow D, Juranek DD. Precautions against biological and chemical terrorism directed at food and water supplies. Public Health Rep 2001;116:3.

[6] Fletcher J, Leach J, Eversole K, Tauxe R. Human pathogens on plants: designing a multidisciplinary strategy for research. Phytopathology 2013;103(4):306–15.

[7] Romeis J, Meissle M, Brunner S, Tschamper D, Winzeler M. Plant biotechnology: research behind fences. Trends Biotechnol 2013;31:222–4.

[8] Broeders SRM, De Keersmaecker SCJ, Roosens NHC. How to deal with the upcoming challenges in GMO detection in food and feed. BioMed Res Int 2012.

[9] Convention IPP. ISPM 5 Glossary of phytosantitary terms. Rome, Italy: Food and Agriculture Organization of the United Nations (FAO) 2007.

[10] Jackson TA, Klein MG. Scarabs as pests: a continuing problem. Coleopts Bull 2006;60: 102–19.

[11] Gubler WD, Smith RJ, Varela LG, Vasquez S, Stapleton JJ, Purcell AH. Pierce's disease UC IPM pest management guidelines: grape online. Davis, CA, USA: University of California, Agriculture and Natural Resources Division; 2008.

[12] United States Department of Agriculture (USDA) Animal Plant Health Inspections Service (APHIS). Movement of plant pests. 7 CFR part 330 subparts 200–212; 2015.

[13] Schaad NW, Frederick RD, Shaw J, Schneider WL, Hickson R, Petrillo 4 MD, et al. Advances in molecular-based diagnostics in meeting crop biosecurity and phytosanitary issues. Annu Rev Phytopathol 2003:305–24.

[14] Han S-W, Sriariyanun M, Lee S-W, Sharma M, Bahar O, Bower Z, et al. Retraction: small protein-mediated quorum sensing in a gram-negative bacterium. PLoS One 2013:8.

[15] Lee S-W, Han S-W, Sririyanum M, Park C-J, Seo Y-S, Ronald PC. Retraction. Science 2013;342(6155):191.

[16] Ronald P. Lab life: the anatomy of a retraction. Sci Am 2013. Available from: http://blogs. scientificamerican.com/food-matters/lab-life-the-anatomy-of-a-retraction/.

[17] Xia R. San Gabriel Valley homeowners swarm to meeting about citrus disease. Los Angeles: Los Angeles Times; 2012.

[18] Xia R, Marcum D. California braces for a deadly stalker of citrus. Los Angeles, CA, USA: Los Angeles Times; 2012.

[19] California Department of Food and Agriculture (CDFA) Citrus disease huanglongbing detected in hacienda heights area of los angeles county. Sacramento, CA: California Department of Food and Agriculture (CDFA); 2012.

[20] Stack JP, Bostock RM, Hammerschmidt R, Jones JB, Luke E. The national plant diagnostic network: partnering to protect plant systems. Plant Dis 2014;98:708–15.

[21] Levy L, Shiel P, Dennis G, Lévesque CA, Clover G, Bennypaul H, et al. Molecular diagnostic techniques and biotechnology in plant biosecurity Gordh G, McKirdy S, editors. The handbook of plant biosecurity. Netherlands: Springer; 2014. p. 207–34.

[22] Kahn RP, Mathur SB. Containment facilities and safeguards for exotic plant pathogens and pests. St. Paul, MN, USA: American Phytopathological Society (APS Press); 1999.

[23] Sundstrom F, Williams J, Van Deynze A, Bradford K. Identity preservation of agricultural commodities. Davis, CA, USA: UCANR Publications; 2002.

[24] Painter JA, Hoekstra RM, Ayers T, Tauxe RV, Braden CR, Angulo FJ, et al. Attribution of foodborne illnesses, hospitalizations, and deaths to food commodities by using outbreak data, United States, 1998–2008. Emerg Infect Dis 2013;19:407.

[25] Miller JK, Herman EM, Jahn M, Bradford KJ. Strategic research, education and policy goals for seed science and crop improvement. Plant Sci 2010;179:645–52.

[26] Peters J, Stoger E. Transgenic crops for the production of recombinant vaccines and antimicrobial antibodies. Hum Vacc 2011;7:367–74.

[27] Salt DE, Smith RD, Raskin I. Phytoremediation. Annu Rev Plant Biol 1998;49:643–68.

[28] Begley S. Plant-based vaccines challenge big pharma for $3 billion flu market Health. New York: Reuters; 2014.

[29] World Health Organization (WHO). Food and Agriculture Organization of the United Nations (FAO). Codex Alimentarius International Food Standards. Guideline for the conduct of food safety assessment of foods derived from recombinant-DNA plants: World Health Organization (WHO), Food and Agriculture Organization of the United Nations (FAO); 2003.

[30] World Health Organization (WHO). Food and Agriculture Organization of the United Nations (FAO). Codex Alimentarius International Food Standards. Principles for the risk analysis of foods derived from modern biotechnology: World Health Organization (WHO), Food and Agriculture Organization of the United Nations (FAO); 2003.

[31] World Health Organization (WHO). Food and Agriculture Organization of the United Nations (FAO). Codex Alimentarius International Food Standards. Guideline for the conduct of food safety assessment of foods produced using recombiant-DNA microorganisms: World Health Organization (WHO), Food and Agriculture Organization of the United Nations (FAO); 2003.

[32] Diversity CoB Cartagena protocol on biosafety. Montreal, Quebec, Canada: Secretariat of the Convention on Biological Diversity; 2000.

[33] Jaffe G. Regulating transgenic crops: a comparative analysis of different regulatory processes. Transgenic Res 2004;13(1):5–19.

[34] Redick T. Coexistence, North American style: regulation and litigation. GM Crops Food 2012;3:60–71.

[35] Kalaitzandonakes N, Kaufman J, Miller D. Potential economic impacts of zero thresholds for unapproved GMOs: the EU case. Food Policy 2014;45:146–57.

[36] United States Office of Science and Technology Policy. Coordinated framework for regulation of biotechnology. June 26, 1986.

[37] National Institutes of Health (NIH) Office of Biotechnology Activities (OBA). NIH guidelines for research involving recombinant or synthetic nucleic acid molecules. 2013.

[38] United States Department of Agriculture (USDA) Animal Plant Health Inspections Service (APHIS). Introduction of organisms and products altered or produced through genetic engineering which are plant pests or which there is reason to believe are plant pests. 7 CFR part 340. May 2, 1997.

[39] United States Food and Drug Administration (FDA). Guidance to industry for foods derived from new plant varieties. 1992.

[40] United States Environmental Protection Agency (EPA). Federal Insecticide, Fungicide and Rodenticide Act (FIFRA). 7 USC Chapter 6. September 28, 2012.

[41] Madden LV, Wheelis M. The threat of plant pathogens as weapons against US crops. Annu Rev Phytopathol 2003;41:155–76.

[42] Gleba Y, Klimyuk V, Marillonnet S. Viral vectors for the expression of proteins in plants. Curr Opin Biotechnol 2007;18:134–41.

[43] Kalaitzandonakes N, Alston JM, Bradford KJ. Compliance costs for regulatory approval of new biotech crops Regulating agricultural biotechnology: economics and policy. Springer; 2006;37:57.

[44] Gaj T, Gersbach CA, Barbas CF. ZFN, TALEN, and CRISPR/Cas-based methods for genome engineering. Trends Biotechnol 2013;31:397–405.

[45] Chen H, Lin Y. Promise and issues of genetically modified crops. Curr opin plant Biol 2013;16:255–60.

[46] Cohen J. Alarm over biosafety blunders. Science 2014;345:247–8.

[47] Tadle M, Henstridge P. Agricultural biosecurity communications and outreach Gordh G, McKirdy S, editors. Handbook of plant biosecurity. Netherlands: Springer; 2014. p. 207–34.

[48] Committee of Sponsoring Organizations of the Treadway Commission (COSO). Enterprise risk management—integrated framework. 2004.

[49] D'Arcy SP, Brogan JC. Enterprise risk management. J Risk Manage Korea 2001;12:207–28.

Select agent program impact on the IBC

Jeffery Adamovicz

Chapter Outline

Introduction and background

The classical role of the IBC

The role of the Institutional Biosafety Committee (IBC) has not been universally defined, and each Institute creates a unique charter for this oversight committee. There are specific IBC guidelines promulgated by the National Institutes of Health (NIH) involving work with recombinant DNA [1]. There have also been recent guidelines issued by the National Biodefense working groups and others on "gain of function" (GoF) experiments and the NIH on Dual use Research of Concern (DURC) [2]. The IBC may be directly involved in approving or referring these experiments to the NIH for approval. Other traditional roles for the IBC include review and approval of all work with infectious agents, work with biologics including nanoparticles, infection

Ensuring National Biosecurity. DOI: http://dx.doi.org/10.1016/B978-0-12-801885-9.00010-X

of animals or plants and/or microbiological work including projects in the classroom laboratory. Broadly, the IBC serves to ensure that biological experiments conducted within their Institute are within regulatory guidelines to protect the Institute from financial and/or legal liabilities and also that laboratory workers conduct these activities as safely as is reasonable in line with current safety recommendations. In other words the primary questions the IBC should ask when reviewing protocols are "Is it safe?" and "Is it legal?" A review of the regulatory requirements of the IBC is covered in Chapters 2, 4, and 5 of this text.

Select agent program

Beginning in 2003 with the expansion of the national biocontainment capabilities in both facilities and personnel, more IBCs had to address new regulations on the so-called select agents [3,4]. Select agents are those bacteria, viruses, and toxins deemed dangerous enough that they require special biosafety and biosecurity precautions and facilities. For instance, they require extra safety training and procedures and background checks to vet the persons working with or with access to these agents. Because of these additional requirements, IBC committees had to expand their capabilities and expertise to adequately address review of protocols involving select agent research. Not all biosafety level 3 (BSL-3) or 4 (BSL-4) pathogens or toxins are select agents but in most cases all select agents require containment in the appropriate BSL3/4 laboratory. When infectious work involves animal models of disease, most select agent studies must be performed in animal biosafety level 3, 3-Ag, or 4 facilities (ABSL-3/3-Ag/4). These studies pose a challenge for IBC oversight as they are not strictly the purview of the Institute of Animal Care and Use Committee (IACUC/ACUC). The purpose of this chapter is to highlight approaches and strategies the IBC can employ to best meets its obligation to review select agent, GoF, and DURC research and ensure the work is properly reviewed so that it is both safe and legal.

Challenges and options

Expanding roles of IBCs: select agent program requirements outside of (or in addition to) NIH OBA

The IBC, and in particular the Biological Safety Officer (BSO), has a myriad of new regulatory compliance issues and other challenges to consider when reviewing/ approving microbiological research involving select agents. In addition to its mandated role in ensuring compliance with the NIH *Guidelines* for research involving recombinant and synthetic nucleic acids, the IBC traditionally reviews microbiological protocols for biological safety, for compliance with CDC guidelines for select agent research, for dual use research of concern (DURC), and for gain-of-function (GoF) research projects. When presented with research protocols involving select agents the IBC must be particularly diligent in assessing whether the work also falls under DURC policies or GoF guidelines [5].

Gain-of-function research

A current area of emphasis for IBCs is to identify and closely monitor biological GoF research. There is interest at the highest levels of government in controlling or regulating this type of research activity. The National Academy of Sciences has recently published a report on GoF research. This report was issued following at least two federally directed research stoppages on GoF agents of concern [6]. Categorically, one could debate the wisdom of these moratoriums on on-going research. However, clearly this needs to be resolved quickly as the negative impacts on acquiring knowledge during a time when these emerging viruses are causing human infections are substantial. While it is important to maintain oversight on this type of research, many believe that the benefits outweigh the risks. While this debate is outside the scope of this chapter, it is clear that IBCs will be increasingly drawn in on the subject of managing GoF research. Currently, the GoF agents of concern include respiratory viruses such as highly pathogenic avian influenza (HPAI), and the corona viruses that cause Middle East respiratory syndrome (MERS) and severe acute respiratory syndrome (SARS). Although there is currently no specific regulatory requirement for the IBC to report GoF research activities, other than those experiments which are classified under current NIH guidelines as "Major Actions," there is an institutional interest in identifying GoF research, in particular for respiratory pathogens. Additionally, specific regulatory guidelines on influenza, SARS, and MERS will likely be promulgated by the federal government in the near future and the list of gain-of-function restricted agents may expand [6]. The underlying concern is that GoF research needs unique biosafety and biosecurity solutions because it leads to the creation of specific pathogens with increased virulence, capability to thwart the immune response or defeat medical countermeasures or that the published knowledge from these studies will enable a terrorist or nation to create biological weapons transmissible through aerosols. Obviously, not all GoF experiments involve this subset of respiratory viruses or other select agents but those that do pose difficult issues for the IBC to manage. Most GoF research in select agents requires approval by not only the institution IBC but also the NIH Recombinant DNA Advisory Committee (RAC). The regulatory concerns may become more complex when a GoF project is also Dual Use Research of Concern (DURC) such as those involving certain noncontemporary or HPAI strains of influenza virus. For a more complete discussion of research that qualifies as DURC, the reader is referred to Chapter 6. The IBC can best serve its institution, the research staff, and itself by articulating clear policies for identifying and reviewing GoF/DURC projects early and streamlining their review. This is optimally done during grant preparation. The PI and IBC would first identify whether a grant contains GoF or DURC research, and develop a risk-mitigation strategy which would be included in both the grant and the IBC protocol. A rubric to illustrate this review process is included in Figure 10.1. To better explain how this process should work from the perspective of the IBC, specific examples of GoF research are highlighted in the following case studies.

Mouse pox

In 2001, a paper was published as a result of studies with an altered ectromelia virus, which is the causative agent of mouse pox [7]. Mouse pox is not a select agent or

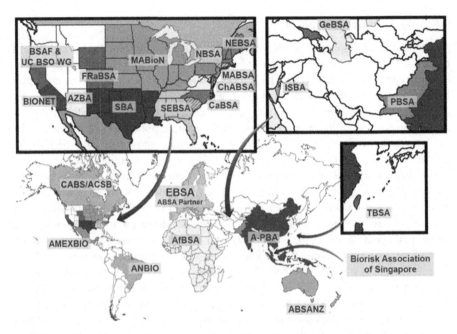

Figure 10.1 Regional and International Biosafety Organizations (http://www.absa.org/
trainingtools.html). The American Biological Safety Association was established as a
US-based organization but over time has formed international relationships with common
goal of improving biosafety. Asian-Pacific Biological Safety Association (A-PBA),
Association of Biosafety for Australia and New Zealand (ABSANZ), African Biological
Safety Association (AfBSA), Association of Mexican Biosafety (AMEXBIO), Association
of National Biosafety (ANBIO), Arizona Biological Safety Association (AZBA), Biological
Safety Information Network (BIONET), Carolinas Biological Safety Association (CaBSA),
Chesapeake Area Biological Safety Association (ChABSA), Canadian Association for
Biological Safety (CABS/ACSB), Front Range Biological Safety Association (FRaBSA),
Georgian Biological Safety Association (GeBSA), Israeli Biological Safety Association
(IBSA), Midwest Area Biological Safety Association (MABioN), Mid-Atlantic Biological
Safety Association (MABSA), Northeast Biological Safety Association (NBSA), New
England Biological Safety Association (NEBSA), Pakistani Biological Safety Association
(PBSA), Southern Biosafety Association (SBA), South East Biological Safety Association
(SEBSA), Taiwanese Biological Safety Association (TBSA). Colors denote the relative
respective geographical areas in which these organizations function. (For interpretation of
the references to colour in this figure legend, the reader is referred to the web version of this
book.)

DURC agent, however because it is a model for human pox virus in the mouse,
research with this agent might raise scientific concerns, especially GoF concerns [8].
The intended purpose of the research was to develop and study a virally vectored
contraceptive vaccine construct including the gene for targeted zaona pellucida glycoprotein 3 (ZP3) which had been previously developed with the addition of the mouse
IL-4 gene. The concept was that IL-4 expression would delay viral clearance and thus

enhance the formation of antibodies and memory cells to the ZP3, leading to an extension of the period of infertility previously observed in ectromelia-ZP3 vaccine studies. However, the observed effect was to make the construct lethal in susceptible mice, and in previously ectromelia-resistant mice. Therefore, this was an unintended GoF experiment. The authors attributed these results to the immunosuppressive effects on cytotoxic T lymphocytes (CTL) and natural killer (NK) cells. Following publication of this manuscript, there was concern that this paper was a road map to creation of weaponized smallpox virus (*Variola major*). While there is ample room to debate this conclusion, the bigger issue is whether either the authors or their IBC should have predicted this result. There was previously published work similar to the Jackson study using vaccinia virus [9,10,11]. These studies did generate data that indicated IL-4 coexpression led to measurable reduction in Th1 cytokines, which are critical for antiviral CTL and NK cell responses. Therefore, a case could be made that the results reported by Jackson should have been foreseen. However, given the differences in pathogenesis of vaccinia and ectromelia in the mouse model, a counter-argument could be made that the highly immunosuppressive results could not have been predicted in the ectromelia study. For the purposes of this discussion it is not important to find fault or assign blame. What is important is to determine what, if any, actions an average IBC could or should take when presented with a similar proposal. Looking at Figure 10.2 we can determine that the Jackson study is not DURC. However, it does appear to fall under what we would now call GoF. Therefore the PI and the IBC would likely need to address these concerns formally before the studies are approved. The authors would then need to create a rationale of why the benefits of the proposed study outweigh the perceived risks. The IBC should be able to articulate their specific concerns perhaps citing the previous studies in vaccinia. Finally, a risk mitigation plan that includes the anticipated results of the study should be proposed, given the legitimate concern over how the information could be misused for altering other pox viruses. The IBC should communicate with the NIH-OBA to determine whether the risk mitigation plan adequately addresses any perceived risk in the proposed studies. This case study may or may not be covered explicitly in new federal GoF regulations but it likely represents the more common type of GoF problem for IBCs. Each institution will likely need to develop its own policy on what constitutes a GoF concern and communicate this policy to their investigators. Additional examples of case studies of this type can be found on the websites for the Federation of American Scientists (FAS) [12] and National Institutes of Health (NIH) [5,13].

Influenza

Future GoF research efforts on influenza virus, particularly highly pathogenic influenza virus, are likely to be affected by new federal oversight, select agent and DURC regulations. Highly pathogenic avian influenza strains are select agents, DURC agents and in some cases GoF agents if they are noncontemporary strains. The initial concern about influenza virus in particular was a result of a published study on creation of an H5NI virus that gained the ability to become transmissible by air between ferrets [15]. The observed increased transmissibility did not lead to increased pathogenicity (though the latter could not have been precisely predicted). The newly created virus

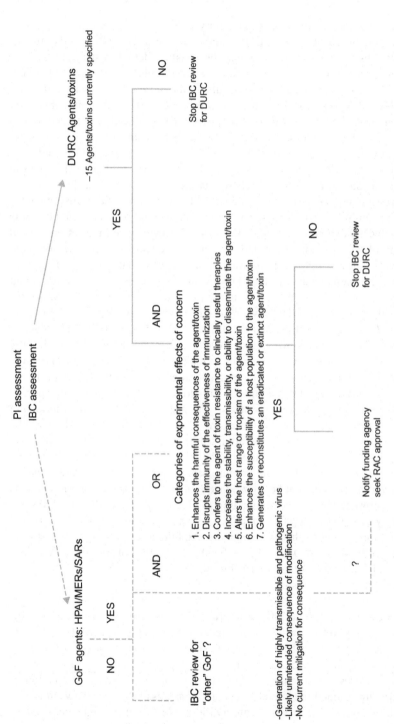

Figure 10.2 Integrated IBC Assessment Methodology for GoF/DURC Grants/Protocols. Categories of experimental effects of concern are specified for DURC protocols. The DURC agent list is available at Ref. [14]. Blue line, current regulatory requirement; yellow, possible regulatory/IBC approach. (For interpretation of the references to colour in this figure legend, the reader is referred to the web version of this book.)

PI assessment
IBC assessment

GoF agents: HPAI/MERs/SARs

NO   YES

IBC review for
"other" GoF ?

AND

OR

AND

YES

DURC Agents/toxins
~15 Agents/toxins currently specified

NO

Stop IBC review
for DURC

Categories of experimental effects of concern

1. Enhances the harmful consequences of the agent/toxin
2. Disrupts immunity of the effectiveness of immunization
3. Confers to the agent of toxin resistance to clinically useful therapies
4. Increases the stability, transmissibility, or ability to disseminate the agent/toxin
5. Alters the host range or tropism of the agent/toxin
6. Enhances the susceptibility of a host population to the agent/toxin
7. Generates or reconstitutes an eradicated or extinct agent/toxin

-Generation of highly transmissible and pathogenic virus
-Likely unintended consequence of modification
-No current mitigation for consequence

?

YES

NO

Notify funding agency
seek RAC approval

Stop IBC review
for DURC

was susceptible to both the currently recommended influenza vaccine and the licensed neurominadase inhibitor Oseltamivir. It appears that the authors of this paper and their respective IBCs exerted more than due diligence in reviewing, approving and monitoring this research project [2]. The concerns raised as a result of this research were initially focused on the inadvertent creation of an influenza strain with pandemic potential. This concern then became more global: creating pathogens that are transmissible via the aerosol route and thus more contagious or able to infect new hosts. The IBC must weigh the value of the information derived from the research study against these and other potential risks. In the case of the Imai study the balance of the equation of risk versus reward seems tilted towards support for the research. The information gained from the Imai study does improve our basic understanding of the influenza virus and in particular key amino acids that affect transmissibility. This information may actually help the public health community better respond to a potential future high-pathogenic avian influenza epidemic. In cases such as this, where the IBC lacks sufficient in-house expertise to address the risk–benefit ratio of a specific protocol or study, they should identify and consult with pathogen-specific experts outside the committee and/or seek additional guidance from NIH-OBA [16].

NIH major action

The NIH currently defines a Major Action as "the deliberate transfer of a drug resistance trait to microorganisms that are not known to acquire the trait naturally, if such acquisition could compromise the ability to control disease agents in humans, veterinary medicine, or agriculture" [1]. Because many recombinant strains are constructed using antibiotic resistance as a selectable marker, the IBC must pay close attention to protocols that propose such developmental experiments. Sometimes the IBC's role in studies is to help avoid having a proposal fall in the Major Action category. Two common methods include demonstration that the antibiotic resistance marker is already present in environmental isolates, or with qualified help, make the argument that a resistance marker is not clinically relevant. In the first case, if a researcher or the IBC can demonstrate that a strain can naturally acquire a resistance marker, then the proposed work may be deemed to not be a "major action." In the latter case the IBC may work with the researcher to prove that there is no resistance to clinically useful drugs. As an example, if a researcher proposes to use kanamycin as a marker in *Burkholderia* studies, a current tier 1 select agent, they could demonstrate that the aminoglycoside kanamycin does not confer resistance to commonly employed aminoglycoside drugs such as gentamicin or streptomycin. This evidence could be obtained from the literature or *de novo* research. If such evidence could be provided for this proposal the research would not be classified as a major action. If, however, the research is deemed to be a major action, the IBC must refer the protocol to NIH-OBA for review and approval.

Responsible officials: dual-duty biosafety officers and responsible officials (biosafety and biosecurity)

In the United States, the CDC and USDA administer current regulations for select agent research, and discussions covering these issues are included in Chapters 1 and 2.

The overarching guidance for biocontainment-based select agent research comes from the manual *Biosafety in Microbiological and Biomedical Laboratories (BMBL)* [17], currently in its fifth edition [17]. This text should be considered a primary guidance document for the IBC when addressing questions related to select agent protocol biosafety review and approval. Elsewhere the WHO guidelines on microbiological research are followed [18]. The institutional BSO is responsible for overseeing biosafety operations in the laboratory and the Responsible Official (RO) is responsible for the security of select agent inventories and overall compliance with the select agent program inclusive of biosafety. These two individuals should be members of the IBC as either voting or nonvoting members. Their primary function during the review is to identify specific issues for consideration by the IBC that are unique to select agent research. These issues may include compliance of laboratory facilities, security concerns, or safety concerns for the intended research. These roles may also be performed by other experienced and knowledgeable IBC members when the RO or BSO are not present or lack the requisite knowledge. The IBC should not rely on an appointed BSO/RO who does not have relevant biosafety experience, including containment laboratory operations and regulatory compliance for select agent research. As an example, in many institutions, BSO duties are often assigned to someone in the Environmental Health and Safety (EH&S) office as an additional duty. Often this might be a chemical hygiene or occupational health expert. However, EH&S professionals will likely lack the requisite biosafety competencies required of the BSO. In such a circumstance, the IBC should seek outside expertise to supplement the committee's review of protocols involving select agent research. Examples of this supplementation are discussed below.

Institutional oversight and IBC quality improvement of select agent research-related issues

Membership and use of subcommittees

Collectively, most would agree that the ideal IBC would consist of highly motivated properly trained and/or experienced individuals that have a mindset focused on assisting the research efforts for safe and compliant research while protecting their institutions from bad publicity, fines, and possible lawsuits. There is no single accepted approach that leads to the development of the right IBC mindset. However, when the IBC focuses on what is optimal for the applicant as distinct from what is optimal for the IBC unnecessary conflicts and delays can be avoided. IBCs that are the most successful at implementing this approach often have direct support from their institutions. This support may include financial assets or direct support in the form of administrators that can implement or assist the IBC members with "best practices." These may include IT-based solutions such as websites with protocols, information products, and/or templates to help the investigator submit their proposal in a timely fashion and identify and address regulatory issues.

A good IBC can often have significant impact and shorten times for IBC approval by becoming involved in research projects, which fall under their oversight as early as possible. Often that time would be during the preparation of the grant proposal. When most grants are prepared, the PI recognizes the need for IBC approval for certain aspects of their work. However, IBC protocol applications are often deferred pending a firm grant award. The IBC may be able to contribute to the substance of a proposal by identifying potential biosafety and biosecurity issues early on, limiting risk to the institution should the grant be funded, and also likely shorten the time-line for IBC approval of the formal protocol. The IBC may form a subcommittee that works closely with their Office of Research to identify and prereview protocols that have identified IBC compliance or select agent research issues as components of the submitted proposal. This subcommittee should work directly with the PI to identify and mitigate issues associated with select agent research.

A second subcommittee that IBC may find useful is one focused on occupational health and worker safety [19]. This committee could consist of one or more members of the occupational health program (OHP), an institute clinical representative, a representative from human resources, and/or an institute legal representative. A health and safety program provides a means to ensure that the risks associated with laboratory activities can be mitigated to best protect the scientific staff and the community. Thus, input from OHP can be a valuable part of the IBC protocol review process. A medical monitoring program, if appropriate, can also help establish inclusion and exclusion criteria for lab staff based on perceived risks and medical conditions of employees for potentially hazardous research. The IBC should work closely with OHP at their institute to recommend specific medical countermeasures matched to the risk involved in each procedure/laboratory. These countermeasures may include vaccinations or plans for provision of therapeutic treatments, such as oseltamivir or ciprofloxacin, should there be a high level of suspicion of the possibility of exposure. This committee may also be charged with helping to prepare agent-specific information documents for the PI, the biocontainment lab's biosafety and incidence response plans and/or for institute or private clinicians or healthcare providers. Finally, this committee can also develop recommendations for initial and/or on-going assessment (physical and/ or emotional) of select agent workers in concert with the RO. This final task is often performed outside the purview of the IBC but inclusion of an IBC subcommittee often closes the loop on several issues including communication of requirements, fostering understanding of the IBC on the rigorous oversight of select agent researchers, and understanding the nature and extent of time and training which is required to obtain select agent clearance from both the federal and institute regulators.

Additional support is available in the form of professional and academic outreach, fellowships, and mentorships. The IBC should ideally have one or more biosafety professionals on the committee. This expertise will help identify and address unique challenges when studies involve select agents or biocontainment activities. If such experience is not available, there are experts in the field and different geographical areas that can be called on for assistance. To ensure successful recruitment of *ad hoc* experts, the IBC should join and/or support regional biosafety organizations. A representation of the ABSA affiliate organizations is shown in Figure 10.2. ABSA and its regional affiliate's membership include a large number of biological safety experts with nationally recognized credentials whose expertise can be brought to bear on problems ranging from regulatory compliance to protocol-specific biosafety/ biosecurity concerns.

Biosafety, biosecurity, and surety expertise are gained by recruiting and utilizing staff and committee members with specialized backgrounds and expertise. In general, the IBC is responsible for ensuring that the committee membership includes individuals with the requisite technical expertise to properly assess the risk involved in proposed studies and whether proposed facilities and procedures are adequate to mitigate these perceived risks. One area which is often overlooked is the inclusion of persons with expertise in biosecurity. These responsibilities are often assigned to the Biological Safety Officer (BSO). However, not all assigned BSOs have either training or practical knowledge to properly perform biosecurity duties for the IBC. A person with nationally recognized credentials, such as a Registered Biosafety Professional (RBP) or Certified Biosafety Professional (CBSP), will usually meet the requisite requirements. The IBC should develop procedures to assess the expertise of their appointed biosafety officer or seek outside expertise to supplement the committee as needed. In addition to the basic requirements in biosafety and biosecurity, two additional areas of knowledge must be covered by the IBC: biological surety and gain-of-function (GoF) research. These areas may be the purview of the BSO or may be covered by other IBC members.

Special surety requirements

Just as NIH regulates rDNA research and requires the IBC to implement policies and oversight to ensure their guidelines are met, other federal agencies such as the Department of Defense (DoD) have special compliance rules. The IBC has a role in compliance of these DoD rules. Institutions that accept grant funding from DoD for research involving select agents must implement a biological surety program. Biosurety is described in Army Regulation 50-1 and DoD directives 5210.88 and 5210.59 [20,21]. These documents describe personal reliability or surety requirements beyond those required by CDC or USDA. The IBC must ensure that these directives are met, for example by establishing working groups or subcommittees with members from the relevant working groups including *inter alia*, the institute research office, human resources, legal, IBC, IACUC, and IRB.

Lack of appropriate biocontainment experience on the IBC

It is not uncommon for members of the IBC to be unfamiliar with noncontainment laboratory space and procedures in their institute and even more likely the case for biocontainment work. An example of a common issue in the nonbiocontainment space that affects safety includes issues such as access to autoclaves. If a laboratory worker has to travel through heavily trafficked public corridors, up or down public elevators, and then has to stage waste outside of the autoclave in a public area, this represents a potential increased biosafety and biosecurity risk as compared to an in-lab autoclave. IBC members unfamiliar with the laboratory layout may not perceive this risk. The same is true for biocontainment operations. Interestingly, a retrospective study which examined the effect of NIH site visits on improving oversight and regulatory compliance for rDNA research by IBCs [22] has not yet been extended to work on biocontainment operations. The IBC can seek one of many outside experts to evaluate their containment laboratory operations and facilities and provide feedback to the IBC. Often IBCs may be hesitant to pursue this route through government regulatory agencies as findings can impact on-going laboratory operations and/or be construed as punitive, given the nature of the findings. There are also professional consultants available for hire to perform this type of assessment but this may be cost-prohibitive. The American Biological Safety Association (ABSA) has proposed a site-visit-based accreditation program [23]. The role of the IBC in the accreditation program is not specified but it would be reasonable to assume they should play a role in compliance and problem resolution following accreditation visits. Because ABSA is not a government regulatory agency and their findings are confidential, this may be a reasonable way to educate the IBC, identify and remedy shortcomings and improve the safety and security of biocontainment laboratories, especially those which conduct select agent research.

Complex projects involving animals

The IBC should establish procedures to address biosafety and biosecurity risks associated with all aspects of select agent work in animal studies. These include, but are not limited to, examination of the waste stream, ensuring the animal room is properly posted with biohazard signage, determination of the infectious risks from animals to humans and if select agents are involved special precautions to prevent theft or misuse. The IBC should develop a pathogen road map to ensure that all biosafety and biosecurity concerns involving select agents are addressed by the PI. An example of this road map is shown in Figure 10.3. In this figure we see the traditional roles of the IBC and IACUC and the areas of new or expanded emphasis the IBC should expect to evaluate for new research proposals involving select agents or select agent studies in animals. The IBC should use the roadmap or a similar template or checklist to determine whether the PI has addressed all critical steps in their protocol. The CDC and USDA often request this type of information from the PI at the time application is made for permission to work with a select agent. Often, the BSO or RO can bring this information directly to the IBC and the IBC should then ascertain whether those procedures are present in sufficient detail to mitigate the identified risks. The institutional Biosafety Officer can help identify and address biosafety and biosecurity gaps for both select and nonselect agents. An effective way to ensure this occurs is to assign the BSO as a member of both the IBC and IACUC committees.

As is often the case in infectious disease research there is significant overlap for safety concerns when animals are intentionally exposed to infectious agents. Both the IBC and the IACUC committees may only perceive portions of the risks associated with such work. Members who have served on both the IBC and IACUC are invaluable in identifying these overlap risks. Risks include biosafety-related procedures for propagating and manipulating select agent cultures, security of select agents, sharps management, animal exposure methods, animal-to-human transmission risks, and overall waste management. A risk commonly overlooked by the IBC and IACUC

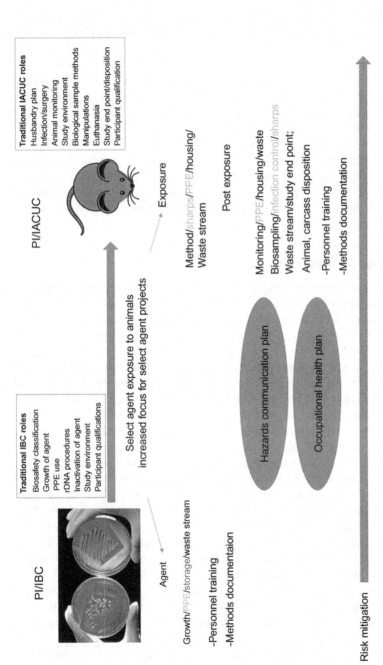

Figure 10.3 Pathogen road map. The traditional roles of the IBC and IACUC are shown as insets. These roles are expanded in scope when overseeing select agent research projects and are more complex when animals are involved. Risk is mitigated by implementing review for all items shown in the graphic, including the development and use of hazards communication tools and enrollment of affected persons in an active occupational health program. Font colors indicate issues involving both biosafety and biosecurity aspects (black), principally biosafety (blue) and biosecurity (red). (For interpretation of the references to colour in this figure legend, the reader is referred to the web version of this book.)

is management of animal waste and bedding. If infectious agents are present in the animal waste they pose a biosafety risk that the IBC and/or IACUC may not perceive or address as they may think the other committee will address the hazard. The IACUC considers allergens a primary exposure of concern, and the issue of animal bedding is often neglected altogether by the IBC. When an animal is infected with a select agent that animal, its biological samples, and often animal waste are classified as select agents. If a select agent can be shed in animal secretions the disposition of contaminated materials must be addressed in the IBC protocol and should also be addressed in the IACUC protocol.

Figure 10.3 also highlights additional risk mitigation tools. First, is the role of an effective occupational health program, enrollment in this program is not required for work with non-tier 1 select agents but is required for tier 1 select agents. Often, the best practice is to enroll all select agent research staff in the occupational health program to fully cover management of occupational exposure to select agents. A second useful tool is the development of a hazards communication plan or tools. Essentially, this tool can be tailored to include the entire laboratory, separate research spaces, or even individuals. The components of this tool/plan include hazard identification methods (e.g., labels, room placards), communications methods (e.g., in person meetings or text messages such as "Anthrax in use in room XX on these dates"), and what needs to happen if something goes wrong (e.g., Incident Response Plan). Finally, it is important to note that it is not the primary role of the IBC to dictate how the PI establishes proper safety and security protocols *per se* but rather to evaluate the adequacy of the procedures to meet the intended purpose, the guidelines established by NIH or the *BMBL* [17] or other specific institutional regulation or best practice. In general, the PI and IBC should work together to identify areas of risk in the research proposal, define appropriate methods and evaluate these methods for adequacy in terms of both biosafety and biosecurity and, finally, propose risk mitigation strategies in cases where the hazard may not be fully understood.

Infrastructure and resource challenges

IBCs develop and implement detailed procedures and policies for users. However, they often overlook the burdens, sometimes unnecessary, that these procedures place on the research community. Often the retort to complaints by PIs about overly onerous requirements is "that's the policy," a response that may create frustration and tension amongst the research staff that in turn may lead to noncompliance. When the IBC takes a different approach and seeks to improve and streamline application procedures, there is often more buy-in from researchers and better compliance. An important factor in compliance is protocol turnaround times. Many laboratory protocols are time-consuming to write and researchers frequently find themselves facing deadlines. Anything the IBC can do to reduce turnaround times on reviews is highly valued and appreciated by the research staff. Recognizing this key item, the Biohazard Compliance Office (BHC) at the University of New Mexico implemented internal procedural changes that reduced their turnaround times from 14.3 to 0.9 days [24]. Almost any IBC can reduce review and turnaround times for IBC protocols with such an introspective approach that values the PI as their customer versus someone who needs to be "regulated." The IBC itself, or the Institute Official, may also wish to use processing times as a measure of efficiency or responsiveness or an indirect measure of workload for the IBC. This may be useful in discussions on administrative or resource support for the IBC.

Biocontainment facilities are unique

The IBC plays a role in not only the initial design and modification of containment laboratories but also in the safe continuing operations of the facility. Therefore, the IBC must have a detailed understanding of their biocontainment facility design and operation. When the IBC is not involved in design or modification discussions a knowledge gap is created. No two biocontainment laboratories are the same in layout though many are similar in function or capabilities. This happens for a number of reasons including architectural design, budget, user input, building restrictions and most importantly the various interpretation of standards for construction. Standards affecting construction of the facilities themselves are often drawn from multiple sources including local building codes, state building codes, and guidelines such as those in the NIH *Guidelines* for construction [25], USDA facility design guidelines [26], *Biosafety in Microbiological and Biomedical Laboratories (BMBL)* [17] or American Society of Heating, Refrigerating and Air Conditioning Engineers (ASHRE) [27]. Another recent standard that affects air handling within biocontainment facilities is ANZI Z9.14 [28] and the IBC can rely on the results of the tests as set forth in this standard as proof that the facility design is safe. As an example of design variations two different architects will read the BMBL requirement "Floors must be slip resistant, impervious to liquids and resistant to chemicals. Consideration should be given to the installation of seamless, sealed, resilient or poured floors, with integral cove bases." These two architects will in turn design two different solutions after consideration of a number of variable factors including cost, durability, and maintenance. This same scenario plays out for all aspects of the laboratories' design including but not limited to doors, anterooms, autoclaves, shower facilities, etc. Consequently, this poses unique challenges and opportunities to managing risk in the facility both for the staff and the IBC. The IBC should add the Biocontainment Facility Director or Manager to the membership of the committee to address facility-related questions. This person can best address the capabilities and limitations of the facility in general or as it pertains to a particular research proposal. This subject is addressed in greater detail in Chapter 4.

Additional importance of time on biocontainment studies

Biocontainment studies are often scheduled well in advance or are scheduled in a tight window for completion at the biocontainment facility. Biocontainment laboratory spaces are often not assigned to single users or even single pathogens. These spaces often must be decontaminated, and reconfigured to support individual research protocols. Therefore, unexpected delays in beginning work are often critical to completion of planned studies, facilities utilization, and reporting to granting agencies. These considerations impact labor and costs beyond the individual PI. The IBC should be aware of these special biocontainment limitations and establish a fast-track approach to review of biocontainment studies or other procedures which will expedite the process. This is not to say that protocol reviews should be rushed or that due process is skipped to accommodate poor planning on the part of the PI. If the IBC publishes their standing committee review schedule and submission deadlines this greatly aides PI compliance. Additionally, if the IBC identifies the need for biocontainment work at the time of grant submission they can work with the PI to identify their expectations on issues for which the PI must prepare information to support an IBC submission.

Select agent PIs as IBC members

Because select agent or biocontainment research in general is more complex than research at lower biosafety levels, the best and most experienced IBC members have often performed research in biocontainment facilities. This experience provides insights into both risk areas as well as reasonable solutions to biocontainment research. If biocontainment researchers have a positive perception of the IBC process and can be convinced of the need for their expertise they are often more willing to participate as an IBC member. These select agent PI members often know and are familiar with other select agent research occurring in the biocontainment lab and can mentor new or junior faculty in meeting IBC expectations for protocols and procedures.

Summary remarks

Work with select agents often engenders an unreasonable amount of fear and apprehension both in the public and within the scientific community. A knowledgeable and well-run IBC is a key asset in allaying worries about research conducted with select agents. In particular, the IBC can institute sound reviews and implement policies that mitigate risk both to the research staff and the community. Additionally, with capable community members as part of the IBC more effective communication with the public can help educate the community on the benefits of select agent research. Staffing the IBC with the "correct" persons can dramatically streamline select agent research, improve biological safety and security and ensure regulatory compliance by answering the two basic questions: Is it safe? And is it legal?

References

[1] Department of Health and Human Services, National Institutes of Health. NIH guidelines for research involving recombinant or synthetic nucleic acid molecules. 2013.
[2] NSABB Recommendations on HPAI Research. Available from: <http://www.nih.gov/about/director/03302012_NSABB_Recommendations.pdf>; 2012.
[3] 7 CFR Part 331 Possession, Use, and Transfer of Select Agents and Toxins.
[4] 42 CFR Part 73 Possession, Use, and Transfer of Select Agents and Toxins.
[5] NIH Guidance on Gain of Function Research. Available from: <http://osp.od.nih.gov/search/site/gain%20of%20function>.
[6] Sharples FR, editor. Potential risks and benefits of gain-of-function research summary of a workshop. Washington, DC: The National Academies Press; 2015.

[7] Jackson RJ, Ramsay AJ, Christensen CD, Beaton S, Hall DF, Ramshaw IA. Expression of mouse interleukin-4 by a recombinant ectromelia virus suppresses cytolytic lymphocyte responses and overcomes genetic resistance to mousepox. J Virol 2001;75(3):1205–10.

[8] Buller MR. Mousepox: a small animal model for biodefense research. Appl Biosaf 2004;9(1):10–19.

[9] Sharma DP, Ramsay AJ, Maguire DJ, Rolph MS, Ramshaw LA. Interleukin-4 mediates down regulation of antiviral cytokine expression and cytotoxic T-lymphocyte responses and exacerbates vaccinia virus infection in vivo. J Virol 1996;70:7103–7.

[10] Andrew ME, Coupar BEH. Biological effects of recombinant vaccinia virus-expressed interleukin-4. Cytokine 1992;4:281–6.

[11] Schwarz EM, Salgame P, Bloom BR. Molecular regulation of human interleukin 2 and T-cell function by interleukin 4. Proc Natl Acad Sci USA 1993;90:7734–8.

[12] Federation of American Scientists Case Studeis on DURC. Available from: <http://fas.org/biosecurity/education/dualuse/index.html>.

[13] NIH Guidance on DURC. Available from: <http://osp.od.nih.gov/office-biotechnology-activities/biosecurity/dual-use-research-concern>.

[14] Shea DA. Oversight of dual-use biological research: the National Science Advisory Board for Biosecurity. Washington, DC: Congressional Research Service; 2007.

[15] Imai M, Watanabe T, Hatta M, Das SC, Ozawa M, Shinya K, et al. Experimental adaptation of an influenza H5 HA confers respiratory droplet transmission to a reassortant H5 HA/H1N1 virus in ferrets. Nature 2012;486(7403):420–8.

[16] HHS Discussion of HPAI Research. Available from: <http://www.phe.gov/s3/dualuse/Documents/funding-hpai-h5n1.pdf>; 2013.

[17] Wilson DE, Chosewood LC, editors. Biosafety in Microbiological and Biomedical Laboratories (BMBL), (5th ed.). U.S. Department of Health and Human Services, 2009. HHS Publication No. (CDC) 21-1112. Available from: <http://www.cdc.gov/biosafety/publications/bmbl5/>.

[18] Laboratory biosafety manual, 3rd ed. In: World Health Organization, editor. Geneva; 2004.

[19] Landon P, Pearl M, Weaver P, Fitch P. Implementation of an occupational health and medical surveilance program at National Biodefence Analysis and Countermeasures Center. Appl Biosaf 2014;19(1):4–10.

[20] Army Regulation 50-1 Biological Surety. In: Army Dot, editor. Washington, DC; 2008.

[21] Department of the Army Pamphlet 385-69 Safety Standards for Microbiological and Biomedical Laboratories. In: Army Dot, editor. Washington, DC; 2009.

[22] Hackney RWJ, Myatt TA, Gilbert KM, Caruso RR, Simon SL. Current trends in Institutional Biosafety Committee Practices. Appl Biosaf 2012;17(1):11–18.

[23] American Biological Safety Association. Available from: <http://www.absa.org/aiahclap.html>.

[24] Muller TB, Stewart DM, Nolte KB. IBC quality improvement. Appl Biosaf 2009;14(2):68–9.

[25] NIH Design Requirements Manual 2015 Update. Design Requiements Manual News to Use. 2015;1(61).

[26] Agriculture USDO. USDA ARS Facilities Design Standards.

[27] ASHRAE Construction Standards. Available from: <http://www.constructionbook.com/store/product/ashrae-standard-622-2013-ventilation-acceptable-indoor-air-quality-in-low-rise-residential-buildings-86176>.

[28] ANSI/ASSE Z9.14 - 2014 Testing and Performance-Verification Methodologies for Ventilation Systems for Biosafety Level 3 (BSL-3) and Animal Biosafety Level 3 (ABSL-3) Facilities. Des Plaines, Illinois: American Society of Safety Engineers; 2014.

Biosecurity challenges for the IBC: an exploration of the roles and responsibilities of Institutional Biosafety Committees in the age of terrorism and biosecurity, now and for the future

11

Ryan N. Burnette and Nancy D. Connell

Chapter Outline

Introduction

Institutional Biosafety Committees (IBCs) serve as a foundational component to biosafety programs at entities that routinely work with infectious agents, toxins,

Ensuring National Biosecurity. DOI: http://dx.doi.org/10.1016/B978-0-12-801885-9.00011-1

recombinant DNA (rDNA), genetically modified organisms (GMO) and other biological products that may pose a risk to human, animal and environmental health. All procedural considerations associated with ongoing biological experimentation are evaluated by the IBC for compliance with safety guidelines and mandates. Since the origins of the IBC model, the focus has primarily remained on biological safety, with little purpose driven toward issues of biosecurity. However, in the years following the 2001 anthrax attacks in the United States, security concerns have necessitated significant changes to the oversight obligations of these committees. Other sections of this book have discussed the composition, function and purpose of IBCs at length; this chapter will focus on how changes in collective attitudes toward biosecurity, as opposed to biosafety, in combination with federal oversight, will likely force the IBCs to evolve and expand their knowledge base and guidance capacity.

Overview of security issues relevant to the IBC

Recombinant DNA

rDNA, the combining of genetic material in juxtapositions not found in nature, is a traditional area of research methodology overseen by the IBCs; these activities – the original target of IBC oversight – have a less direct relationship to security issues. The use of rDNA techniques in the production of therapeutic molecules (insulin, interleukins, hormones, etc.) is generally well accepted; the use of genetic modification in environmentally (e.g. bioremediation) or agriculturally (GMO) important arenas remains controversial in segments of society, as is the use of rDNA in the treatment of human disease, or gene therapy, despite a number of successes (e.g. alipogene tiparvocec, trade name Glybera, for the treatment of severe pancreatitis caused by lipoprotein lipase deficiency) [1]. However, rDNA techniques are often used in studies of virulence mechanisms, e.g. in the generation of chimeric bacteria and viruses.

Dual use research of concern

The rapid and global advancement of scientific technology in the past two decades is the stimulus for heightened concern with safety issues in several biomedical areas, especially infectious disease, immunology and neuroscience. Scientists are now able with ease to modify infectious agents for the purpose of understanding pathogenesis and virulence; these capabilities and the materials associated with them are easily accessible and no longer require extensive training. Modulation of the immune system to mitigate disease or boost the immune response is now within reach in the therapeutic arena. Imaging and other technologies have enhanced our understanding not only of how the brain works – memory, sensory and motor control, neural networks, etc. – but also how to control neural activity and behavior. In the face of many remarkable achievements in these and other areas of biology, there has been a concerted effort by policymakers, ethicists and scientists themselves to address the dual nature of this kind of work, and specifically, the concept of "dual use research of concern" (DURC) [2]. Biological agents that are considered DURC are those agents that have merit in basic and applied biomedical research, but can also potentially be manipulated and employed for nefarious purposes. Typically, these are agents under active investigation for the development of therapeutic treatments and cures. An NAS report released in 2004 summed up the dilemma [3]:

> *The great achievements of molecular biology and genetics over the last 50*
> *years have produced advances in agriculture and industrial process and have*
> *revolutionized the practice of medicine. The very technologies that fueled these*
> *benefits to society, however, pose a potential risk as well – the possibility that these*
> *technologies could also be used to create the next generation of biological weapon.*

Select Agent program

In addition to DURC, the implementation of the Select Agent program (SAP) in the United States in the wake of 9-11 and the anthrax attacks has placed considerable increased burden on institutions wishing to pursue work utilizing certain microorganisms and toxins. While the details of the SAP and its impact on IBC function are explored in Chapter 10 of this volume and elsewhere, the national security goals of the SAP are clear: to oversee the routes of access to agents that have the potential for malignant use. The US government has designated certain agents as Biological Select Agents and Toxins (BSAT) if they have the potential for widespread negative impact on human, animal and/or environmental health if improperly used, either unintentionally or nefariously. Shipping, storage and manipulation of BSAT require the registration with and approval of the Centers for Disease Control and Prevention (CDC) and/ or the US Department of Agriculture (USDA). There are a total of 29 agents on the CDC's list of BSAT, and 14 overseen by the USDA; 11 are considered "crossover" agents and are regulated by both federal entities. Institutions that maintain BSAT registrations must have programs and facilities that meet specific criteria prescribed by the corresponding agency, overseen by the Division of Select Agents and Toxins, either through the CDC or the USDA/Animal and Plant Health Inspection Service (APHIS). In the United States, most BSAT agents can be acquired for no charge from BEI Resources Repository under the auspices of the NIH; registration and documentation of adequate biosafety oversight (CDC and/or USDA) is required for access to these materials. Accordingly, access control, agent inventory management and personnel suitability assessment come into play. From the perspective of the IBC, members must be aware of both DURC and BSAT.

- Access control: physical security is an essential component of the SAP. Access controls include basic security devices such as locks and keys to more elaborate systems such as card keys and biometrics. These systems allow the documentation of all entries and storage of data for future referral if necessary.
- Agent inventory management: each entity must develop a system to record all transfers into, within and outwith the institution. Furthermore, all activities involving the agents must be recorded and managed. The select agent regulations require an accurate and current inventory for: (i) "each select agent (including viral genetic elements, recombinant and/or

synthetic nucleic acids, and recombinant and/or synthetic organisms) held in long-term storage (placement in a system designed to ensure viability for future use, such as in a freezer or lyophilized materials)"; (ii) "any animals or plants intentionally or accidentally exposed to or infected with a select agent (including number and species, location, and appropriate disposition)" and (iii) "for each toxin held." [4]. The requirement for monthly accounting of every sample remains controversial, with many scientists questioning whether such accounting is at all useful [5].

• Personnel suitability assessment: workers with access to Select Agents are required to undergo detailed background checks. While each entity may design its own system of personnel suitability, the NSABB has defined as suitable a person who is free of felony convictions; has no domestic or international terrorist ties; has no history of scientific or professional misconduct in the workplace; possesses emotional stability and capacity for sound judgment; has a positive attitude toward safety and security measures, and standard operating procedures (SOP); and is free of vulnerability to coercion [6].

Overview of current IBC roles

The IBC in a traditional academic research and/or medical center environment convenes with some frequency or regularity to review protocols submitted by investigators. The requirements for protocol submission may be driven by regulatory compliance, research funding or institutional best-practices. Regardless of the reason, members of the IBC will review the protocol presented, taking into consideration the nature of the biological agent, how it will be manipulated, for what purpose, and evaluate the overall safety. In a way, this is analogous to performing a risk assessment, a hallmark of biosafety programs.

Risk assessments are driven by consideration of two major fundamental parameters: the nature of the biological agent being manipulated and the specific method(s) by which that manipulation takes place (including location and equipment). Simply put, a risk assessment applies qualitative measures to judge the hazards of a biological agent, in terms of its potential adverse effects, taking into account the environment and techniques that are employed in its manipulation. The outcome of a robust risk assessment process is a series of mitigation tactics that can be employed to ultimately reduce exposure of the laboratory, the healthcare worker and the surrounding environment to the biological agent. From the perspective of the IBC, each time an experimental or clinical protocol is reviewed, a risk assessment is being performed. While complete elimination of risk is not possible, this process aims to reduce the risk to its theoretical minimum.

Expertise required

The brief review above of the kinds of research overseen by the IBC shows that the committee membership of IBCs must provide the scientific expertise required to evaluate research protocols and ensure the safety of workers and the public. In addition, the committee can also play an important role in mediating controversies, serving as liaison between the scientific researchers and the public. The NIH Office of Biotechnology Activities OBA provides for such interactions to take place by ensuring that IBC meetings (Section IV-B-2-a(6)) and minutes (Section IV-B-2-a(7)) are open to the public:

> Section IV-B-2-(6): When possible and consistent with protection of privacy and proprietary interests, the institution is encouraged to open its Institutional Biosafety Committee meetings to the public.
> Section IV-B-2-(7): Upon request, the institution shall make available to the public all Institutional Biosafety Committee meeting minutes and any documents submitted to or received from funding agencies which the latter are required to make available to the public.

Security scenario

The IBC may be confronted with complicated scenarios that combine more than one oversight issue and require security analysis. A timely example of an intersection of DURC and BSAT is illustrated in studies of Ebola virus (EBV), the causative agent of a hemorrhagic fever (Ebola virus disease – EVD) that has captured worldwide attention following the outbreaks beginning in 2014 in West Africa. There are no approved therapies or vaccines for EVD, and EVD is the subject of much biomedical research [7]. However, the fact that EVD has no cure and has a high mortality rate makes the causative agent – EBV – attractive as a potential biological weapon, and so the virus is considered an organism with dual use potential. EBV itself is a Select Agent and is designated for manipulation at Biosafety level (BSL) 4, the highest level of containment, available at only a limited number of laboratories in the United States and around the world. However, BEI, the NIH-affiliated materials resource, offers EBV material – specifically, Ebola virus RNA – for experimental use. The nucleic acid carried by EBV is negative strand – designated (−) RNA, which is completely noninfectious. (−) RNA itself carries no direct information capable of forming viable virus particles, but requires an additional enzyme – RNA-dependent RNA polymerase – to be converted to a form that can be directly recognized by the infected cell. Thus, the "neutral" (−) RNA can be manipulated at BSL1, the lowest level of biological containment. This incongruity – that the nucleic acid material is BSL1 while the whole virus is restricted to BSL4 – might set up a difficult decision for an IBC with respect to biosafety level assignment and security measures. There are a number of alternatives that an IBC could consider:

1. that the material can be used in a traditional (noncontainment, BSL1) laboratory;
2. that the material can be used at BSL1 with extra security measures, such as a locked freezer in a locked room with access limited to a small number of workers;
3. that the material is not a biosafety risk *per se* but may pose a high enough biosecurity risk that the material should be used only at BSL3;
4. that the material has a finite risk of triggering the production of infectious particles under the right conditions, and should be used only at BSL4.

Clearly, the combined expertise of the IBC must be brought to bear in determining the combined safety and security measures appropriate for this kind of complicated project. Entities and their IBCs may vary in the final decision on a question such as this.

IBC members must also be knowledgeable and properly staffed to evaluate research and clinical trials that interface with other regulatory requirements and sources of research funding. Human gene transfer is an emerging area of therapeutic value and great investment. In many cases, the genetically manipulated material, such as rDNA, that is used as the therapeutic product was created in a research setting that received funding from the National Institutes of Health (NIH). For those institutions conducting research or involved in clinical trials IBC oversight is a requirement. This also brings yet another possible area in which IBCs must be responsible – the interface with Institutional Review Boards (IRB).

IRB, independent review boards for human subject protection in clinical trials, and Institutional Animal Care and Use Committees (IACUC), boards for the review of ethical animal research, are two other oversight committees with which IBCs frequently interface. Since research involving biosafety and animals or humans, in clinical trials, often overlaps, IBC members are required to understand the interrelationships of these other committees and the nature of their realms of oversight.

It is clear that IBCs have a great deal of information and responsibility serving to support safe research. However, despite all of the duties, responsibilities, expectations and roles that the IBC plays, biosecurity is a concept still early in development in the current IBC framework. As this chapter will demonstrate, it is not unreasonable to assume that the IBCs may be obliged to take on the unofficial role of Institutional "Biosecurity" Committee.

Defining biosecurity

Biosafety is commonly described as a set of practices and principles used to reduce the risk of exposure to a potentially hazardous biological agent or toxin. Biosecurity is also a set of practices and principles, and both overlaps with and is distinct from biosafety. At its simplest, biosecurity is aimed at reducing threat, whereas biosafety is aimed at reducing risk. Those threats can be from within an institution housing and manipulating biohazardous agents (e.g. insider threat), or from an individual from the community who wishes to gain access to the agents (e.g. outsider threat). In both cases, biosecurity is a code of conduct established to reduce the threat of both unintentional and intentional access to biohazardous materials, or valuable biological materials (VBM; e.g. valuable vaccines).

Biosecurity is not a singular program or function, rather it comprises multiple layers that deal with physical security, operational security, personnel reliability and accountability, access control and the mitigation of threats.

Physical biosecurity describes the central dogma of "gates, guards and guns," but in a laboratory setting more subtle measures are usually employed. These measures may include closed-circuit cameras, viewing panes in laboratory doorways and simple locks on doors and storeroom areas. Physical biosecurity also overlaps with *access control*, where keycards or biometrics may be employed restricting access to certain personnel. All of these measures are aimed at reducing the risk of an outside threat.

Operational biosecurity comprises all of the policies, SOPs, emergency response procedures, and manuals that describe how a biosecurity program will work for a given institution. Analogous to a biosafety manual, a biosecurity manual contains all of the necessary documentation required to facilitate management of a biosecurity program.

Personnel reliability and accountability regulates who is granted access to the biohazardous or VBM. The importance of these programs is underscored by the 2001 anthrax attacks, when an employee of a federal research lab is alleged to have used anthrax in a series of biological attacks [8]. This is an example of an insider threat. Personnel intending to use BSAT must receive clearance from the Federal Bureau of Investigation. This background check is one example of the personnel reliability and accountability program. In general, the goal is to take measures ensuring that personnel with access do not pose a threat to other members of the public.

All of these combine to form a comprehensive biosecurity program; yet a review of these concepts may not immediately seem to be an area of concern for IBCs. However, the overlap of biosafety and biosecurity will undoubtedly impart additional responsibilities onto the IBC. The potential paradigm shift presented here is that traditional IBC roles are directed at evaluating and mitigating *risk*; biosecurity, as described, indicates a different perspective – evaluating and mitigating *threats*.

Operational overlap of biosafety and biosecurity

Laboratories carry out a myriad of activities: basic and applied research, epidemiology, sample analysis and pharmaceutical development [9]. All of these activities contribute to the welfare of society; society expects these activities to be carried out securely and safely. On occasion, and despite international instruments designed to prevent such activities, the fruits of biological laboratories have been used for harm. One of the more recent and impactful events is the distribution of highly infectious anthrax spores through the US postal service in 2001; while the FBI has closed the case, the origin of the material used in the attacks remains unknown. Many of the federal guidelines and regulations in the United States, EU and other states were developed and legislated in response to the events that unfolded almost 15 years ago. In the distant past, experimental work in the military research laboratories of the United States, the United Kingdom, the Soviet Union and Japan before and during WWII was carried out with the direct intent of creating biological weapons; other state-level programs have been described [10]. The Biological Weapons Convention (BWC) entered into force in 1975; its primary concern at the time was state-sponsored programs [11]. Since 2001, however, the major focus of global efforts to limit the development of bioweapons, even at the international treaty level [12], is on small group/individual actor activity, i.e. bioterrorism. The safeguarding of dangerous biological agents is the goal for all institutions carrying out the work with dangerous agents – whether research or clinical. Good biosafety practices can provide the groundwork for adequate biosecurity: review of experimental procedures, responsible inventory, identification and vetting of personnel, record of material transfers, and proper treatment of waste. Each of these procedures is essential for safety in a biological laboratory; each of them if carried out correctly will limit illegitimate access to dangerous agents. To complete the list of biosafety and biosecurity practices, we would add investigative protocols and a reporting structure. Indeed, the WHO biosecurity guidelines suggest that an important institutional practice is oversight by the biosafety committee [9].

Rationale for IBC as the unofficial biosecurity committee

The IBC is a common entity at many research institutions. By contrast, independent biosecurity committees are less common. Given the overlap of biosafety and biosecurity, and the lack of independent biosecurity committees, it stands to reason that IBC may adopt the unofficial role of biosecurity committees. There are several observations that support this hypothesis:

- Unlike IRB and IACUC, there is little regulatory framework that dictates the presence and function of IBC or biosecurity committees. A number of recent lapses at federal labs may result in increased scrutiny and subsequent legislation aimed at fortification of IBC [13]. Yet there is virtually no regulatory framework in place to dictate the requirement for a biosecurity committee.
- Biosafety has been codified over a period of decades through guidelines (e.g. *Biosafety in Microbiological and Biomedical Laboratories (BMBL)*, CDC/NIH), resulting in semi-prescriptive foundations for IBCs. In short, biosafety is a familiar concept [15]. No such guidance document for biosecurity currently exists, and no obvious federal agency has purview over biosecurity. In fact, biosecurity was not included in the *BMBL* until the most recent edition published in 2007.
- Resource limitations are a way of life at many academic institutions. Participation on an IBC is often seen as an additional job function with little incentive. IBCs are routinely staffed with members of the research community with expertise in the science as it relates to safety, members of environmental health and safety offices, biosafety officers and community members. Few have the resources to borrow from academic security/public safety programs.
- The level of concern surrounding biocontainment laboratory operations is increasing. It is likely that institutions will attempt to fortify their overall biosafety programs, and this may result in the discovery of a lack of biosecurity. This perceived immediate need to close that gap, combined with a lack of resources, may drive responsibilities related to biosecurity oversight to those that have purview over biosafety programs. In short, there is pressure from federal agencies and the public to "tighten up" biosafety and biosecurity programs.
- Academic security programs are not focused on biocontainment operations; they usually are focused on campus safety and security. This may overlap with campus police forces. At many institutions, there is little connectivity between academic security programs and the IBC.
- Educational and credentialing programs exist for biosafety, but not biosecurity. Biosafety officers commonly have the designation of Registered Biosafety Professional or Certified Biological Safety Professional. Currently there are no such curricula at large for the creation of biosecurity professionals. Again, the convergence of biosafety and biosecurity allows many programs to infer that biosafety professionals also possess biosecurity expertise; this is not necessarily true.
- The need for biosecurity is growing faster than the needed resources. In other words, the job needs to fall to someone, and for the reasons described herein, the IBC may simply seem like a natural reservoir of personnel and related expertise.

All of these observations point to the fact that, while limited, biosecurity expertise may be an implied ideal within the IBC. It is possible that certain academic institutions will elect to adopt and form independent biosecurity committees, or something similar to accomplish the objectives. In many private R&D organizations and biotechnology/biopharmaceutical companies, resources for (bio)security are, at least in part, geared toward the protection of their products, or "value-based management". However, the vast majority of academic programs lack the resources and expertise to implement a robust biosecurity committee or program. While some may question the need for biosecurity programs, it is notable that biosecurity gaps have been reported at many institutions [13]. Taken together, the question is not *whether* colleges and universities need biosecurity programs, it is *how* and *when* they will be implemented.

Strategies to fortify IBC to embrace biosecurity

It is possible that the future of academic research programs may include independent biosecurity committees, with sufficient crosstalk amongst other committees, such as IBC and IACUC, resulting in a comprehensive network that addresses ethics, safety and security. Until resources and expertise are readily available, colleges and universities should consider taking progressive steps to fortifying their IBC to address the immediate needs for biosecurity. The following is a series of questions that address some viable strategies that the IBC can embrace to accomplish this goal.

How can institutions quench the growing concerns of the public? There is a requirement for every IBC to appoint two members of the community not affiliated with the institution. These members might provide some insight into the specific concerns of the community. Working with the institution's media relations department, IBC members can learn how to explain the nature of the work performed by researchers as well as the oversight mechanisms in place.

Can biosecurity be added to the charter of the IBC without imparting undue burden? There has been a gradual accumulation of regulatory requirements for both the researchers and the IBC staff [14]. As an illustration, an investigator planning to work with *Bacillus anthracis*, the causative agent of anthrax, in the year 2000 would have had to ensure that BSL2 practices were in place, register the lab with CDC for transfer of *B. anthracis* cultures, submit protocol to the IBC, consider using a BSL3 laboratory (not required), and perform the experiment. In 2014, the following steps in addition to those above are required: reject possibility of using a BSL2 laboratory, find a Select Agent lab with "hardened design," register lab for possession, use and transfer of Select Agents, obtain permits for strain possession and for shipping, hire two people to do one experiment (the "buddy" system), clear personnel through the institution's Personnel Suitability Program (criminal and psychological background checks), train personnel in experimental methods, biosafety and biosecurity protocols, personal protective equipment use, train and exercise workers nearby in emergency response, submit experimental protocol to the IBC, perform risk assessment of detailed protocol and enter the lab 6–12 months after initiation of project. Clearly, Select Agent research has become extremely complicated and requires the integration of multiple entities across the institution, state and federal government. One might argue that the addition of biosecurity to the charter of IBC would be burdensome only in that additional training of IBC members might require extra commitment; the primary burden of biosecurity requirements already in place lies squarely on the shoulders of the researchers and relevant administrative elements of the institution.

Adapting/expanding the expertise of the IBC to address biosecurity

IBCs, in theory, receive regular training on IBC member roles, responsibilities, pertinent regulations and guidelines, and other topics necessary for efficiency and compliance. Like so many aspects of biological safety and security, training is necessary to provide researchers, leadership and committees information and confidence in subject areas. It is not unreasonable that providing the IBC with biosecurity training would be a valuable initial step to promoting a foundational level of biosecurity expertise.

Training can be provided in many forms, from in-person, instructor-led training, to on-demand webinar training. The most appropriate training vehicle should be established based upon the goals and anticipated outcomes of the training programs. It is likely that the IBC is well-versed in the biosafety program at its corresponding institution, having somewhat detailed knowledge of the biological agents in use, the number and biosafety level of the various laboratories, and a working knowledge of DURC and the SAP. With this background, the use of generic biosecurity training is not as valuable as site-customized biosecurity training that can be woven into the areas with which the IBC members are currently familiar. This will dramatically increase the relevance of the training exercise. Further, instructor-led training tends to be more effective than on-demand webinar training. Regardless, the institution and the IBC must make the choice of the type of training it wishes to receive, based on the goals of the training program itself. For example, it is unlikely that the IBC will ever be involved in the selection and placement of security equipment (e.g., CCTV cameras) so a detailed understanding of physical biosecurity is probably less relevant to the IBC; these kinds of decisions are frequently made by contracted security companies in conjunction with institutional biosafety officers during construction design. In contrast, the biosecurity requirements specified under the Division of Select Agents & Toxins may be extremely relevant to institutions with active SAP registrations.

Relevance in training is key, but what topics related to biosecurity are relevant to train an IBC? Again, weaving in elements that the IBC is already familiar with will increase the efficacy of the training program. Each institution should assess its own "touch points" for biosecurity and develop a list of training topics. At a minimum, academic institutions should consider the following as a potential core list of biosecurity topics worthy of IBC training:

- definition and overview of biosecurity,
- how biosecurity in embedded in biosafety programs,
- importance of access control,
- overview of personal reliability and accountability,
- overview of threat and vulnerability assessments (as a correlate of biological risk assessments),

- Select Agents,
- DURC,
- relevant and pending legislation on biosecurity.

These are topics that are at the core of the mission of the IBC whether or not the IBC becomes the unofficial (or official) reservoir of biosecurity oversight. In addition to these suggested topics, each institution should take the opportunity to pinpoint other topics that may be specific to its own IBC. For example, IBC protocol submission forms may be modified to include a section on biosecurity (discussed below) to be completed by investigators, as part of the IBC approval process. In this case, the IBC should be well-educated on (i) why this section is included in the protocol submission form, (ii) what dictates acceptable completion of this section of the form, (iii) what recommendations to make to investigators if this section of the form is not sufficiently addressed, and (iv) the rationale for the specific composition of this section of the form.

One common gap at many academic institutions is lack of biosafety training for principal investigators, resulting in a disconnect between the IBC and investigators. This is most evident when investigators submit research proposals for IBC approval and biosafety issues have not been appropriately or accurately addressed (e.g., not assigning the biohazardous agent of intended use to the proper biosafety level). Therefore, by extension, training on biosecurity should not be limited to the IBC. Rather, it may become important to provide training to those investigators whose research interfaces with biosecurity issues. This may be those investigators at BSL3, working with Select Agents, or who have known DURC issues. Institutions will have to decide, again, to what level to train investigators, but it is possible to add biosecurity training topics to other training curricula that investigators may already be obligated to complete. For example, all investigators (and students) who work at BSL3 should have the appropriate training, both classroom and "hands-on." The addition of a biosecurity module to the BSL3 training curriculum would prevent the addition of a separate training program, and increase the relevance of the biosecurity topics as they relate to BSL3 operations.

Modification of IBC charter and protocol submission forms

The IBC charter is a foundational element of the IBC, providing the purpose and basic "rules." It usually includes directives on the purpose of the IBC, how membership is achieved and maintained, responsibilities, disclaimer on conflicts of interests, and standard procedures the IBC will follow. In short, this is the anchor document on which IBCs are established and maintained. While this is a living document, and one that should be reviewed periodically, potentially changing the direction of the IBC to include biosecurity is a significant undertaking.

Obviously, the IBC must first decide to what level it will engage in issues pertaining to biosecurity at its institution. An institution that does not (i) participate in the SAP, (ii) has a documented background of no DURC, (iii) does not work with biohazardous agents about Risk Group 2 and (iv) does not operate above BSL2 may elect that there is no need for involving itself or its IBC in any biosecurity-related endeavors. For institutions that do meet higher criteria (e.g., SAP, DURC, BSL2+, Risk Group 2+), it is recommended that some policy to address biosecurity is adopted, and potentially the involvement of the IBC should be considered.

For institutions with a dedicated (bio)security office/committee, the level of IBC involvement may be quite low. In this instance, the IBC may wish to play a supporting role, perhaps having a representative from the (bio)security office/committee sit in on IBC meetings. It may also be mutually advantageous for a member of the IBC to participate in (bio)security committee meetings. This type of relationship is similar to a common relationship between the IBC and offices of environmental, health and safety, and can be critical to a comprehensive understanding of the interplay between health, safety and security initiatives at an institution.

The area of greatest concern, as this chapter has borne out, is those institutions that have determined a stake in biosecurity, but do not possess a dedicated (bio)security office or committee. It is this instance where emphasis on biosecurity may tasked to the IBC. Any specific directives regarding how the IBC will provide oversight on biosecurity issues is likely to come from upper management (e.g., Vice President for Research), rather than self-elected to "police" biosecurity issues. If agreement is reached between upper management and the IBC that the IBC will have oversight of biosecurity issues, then the IBC Charter should reflect that stance. For example, the purpose of the IBC Charter may be amended to reflect the following additions:

The IBC will review and approve:

- *research involving the use of rDNA, as it pertains to the NIH Guidelines,*
- *microorganisms and viruses that are pathogenic to humans, animals and plants,*
- *Select Agents,*
- *DURC,*
- *institutional biosecurity.*

The simple addition to the IBC charter to include *DURC* was a concerted step toward providing biosecurity oversight. These additions can be further defined in the responsibilities and procedures of the IBC Charter.

One method to provide oversight in this capacity is the modification of the IBC, or biosafety, protocol submission form. This is completed by principal investigators conducting research directly related to the topics above, and reviewed by the IBC. The primary purpose of this review is to ensure safety in research practices, and compliance with relevant federal, state and local laws and guidelines. These protocol submission forms typically call for an overview of the research, biological agents proposed and their corresponding risk group, biosafety level at which the work will be conducted, and, as appropriate, specification on the use of rDNA, the use of human tissues or cell lines, and other items, all of which must be addressed by the investigator. Correctly completed, this information allows the IBC to review the proposed research and approve or disapprove of the initiation of research. In conjunction with the proper biosecurity training (discussed above), the addition of a brief section on biosecurity to this protocol submission form may help streamline the decision-making process for the IBC. Here, this section of the submission form may request specific information, such as:

- Select Agent use (and appropriate clearance documentation to work with Select Agents),
- DURC,

- access control measures where research will be conducted,
- researchers' completion of biosecurity training.

Coupled with biosecurity training for the IBC, this may become an efficient tool allowing the IBC to make approval decisions on proposed research in the context of relevant biosecurity issues.

Challenges in applying biosecurity to research programs

Recognition by upper management

Academic institutions usually have various silos of authority that dictate where the IBC falls with respect to operational oversight of the IBC. For example, many institutions have an office of research compliance and integrity. This office oversees not only the IBC, but may also oversee other committees such as IRB and IACUC. These committees may report to Institutional Officials (IOs), who are normally the equivalent of a Vice President for Research, or a Director of Research Compliance and/or Integrity. Despite differences in reporting structure at academic institutions, the IBC is largely staffed and driven by research-related entities within the institution. Therefore, the leaders of academic research programs and/or offices provide the fundamental top-down support of the IBC directives and functions.

Support from the upper level of research is critical for the success of properly administrated IBCs. Therefore, the inclusion of biosecurity objectives into the overall mission and scope of the IBC is something that must be reinforced from upper management. Upper management must embrace the importance of biosecurity, as it relates to the health of the research programs, protection of the research and researchers, and the potential impacts to research funding. No IO desires to be at the helm of a research enterprise that has suffered a breach in biosecurity, such as a disgruntled student who knowingly removes restricted biohazardous materials from a controlled laboratory area. The results of such an event can have lasting consequences to the institution with respect to funding and reputation. Therefore, it behooves upper management to have a strong working knowledge of biosecurity issues that are relevant to the research being conducted, and recognize the contributions the IBC can make to ensure the security of the research programs.

Upper management, like both IBC members and principal investigators, may require some level of biosecurity training. Such training would increase their proficiency in the area and provide a foundation of knowledge to support the IBC. While responsible conduct of research training – that in many cases includes DURC training – is offered to students and other trainees, upper management and investigators are often left out of this loop. Frequent attendance at IBC meetings by upper management is always a good indicator to the IBC that they have the support and commitment from research leaders in their activities. These, among others, are good steps that upper management can take to (i) demonstrate their commitment to the IBC and (ii) protect the security of the research at their institution.

Creating a biosecurity culture without over-policing

One of the more difficult, qualitative objectives to meet is the development of a good safety and security culture at a university. This is not something that can be quantified, or addressed with a checklist, or is codified in federal guidelines. Rather, a good safety and security culture is something that is achieved as a result of strong programs, with top-down support, focused education, strong relationships between investigators and regulatory oversight, and a research-friendly environment.

A common dilemma faced at many academic institutions is a feeling that researchers are "policed" rather than treated as collaborators. Research funding, supplied by world-class investigators, is the lifeblood of academic research institutions. For this reason alone, the health of the research enterprise is a direct correlate to the overall health of the institution. Why, then, would a greater emphasis not be placed on building strong relationships between those that oversee safety and security with those investigators that supply the programs and funding? Often, it comes from a simple element of fear: fear that research efforts are not being conducted safely or in compliance with regulations. This fear often results in a natural distrust of investigators. Admittedly, every research enterprise has its share of troublesome investigators that seem to ignore the rules of safety and compliance.

In the case of biosafety programs, a key element to a good biosafety culture is the personnel that oversee the biosafety program, such as the Biosafety Officer and IBC Chair. These individuals walk between the roles of friend of research and compliance officer. Knowing that resources for biosafety programs are often limited, it is important for these personnel to have trusting relationships with their investigators. This is often achieved with a simple "open door" policy that invites investigators to reach to the biosafety office and IBC for help and support. Research institutions with personnel that conduct regular training, frequent laboratory visits, invite the participation of investigators with the biosafety office and IBC, are usually those that have a substantive biosafety culture.

The integration of biosecurity into research programs will likely face many of the obstacles and growing pains that biosafety programs have encountered. However, a good biosecurity culture will be more easily introduced and maintained at those institutions with a strong biosafety program. This is because biosecurity (i) is embedded into biosafety and (ii) is integrated as an additional element, rather than an entirely new, separate program. The IBC can play a pivotal role in the adoption and incorporation of biosecurity into existing biosafety programs.

While the IBC is a research oversight committee, it does have some responsibility for bridging the divide between investigators and compliance issues. An IBC that welcomes investigators who are submitting protocols for review to IBC meetings will provide those investigators an opportunity to discuss the proposed research and educate the IBC. This simple, proactive step will assist in the development of a scrupulous biosafety culture. Likewise, if the IBC becomes the institution's reservoir of biosecurity oversight, investigators should be encouraged to discuss biosecurity-related matters with the IBC.

As discussed earlier, seeding members of the IBC onto other committees (e.g., campus security) and reciprocal participation on the IBC from members of other committees, will result in cross-pollination of biosecurity ideas, and breed a collaborative environment.

A path forward: a "BMBL" for biosecurity

There is little doubt that the process of integrating biosecurity principles and practices into the IBC, and thereby the overall research institutions, begins with proper advocacy, education and training. Biosecurity, as compared to its counterpart, biosafety, is still developing. There is little guidance on how to implement biosecurity programs at academic research institutions. In fact, it was the 5th edition of the *BMBL* released in 2009 that contained the first section on biosecurity. To date, there is no equivalent document for biosecurity. For universities, biosecurity rarely exists outside of the SAP. Like biosafety, biosecurity is not a "one size fits all" set of practices and principles. It must be scaled appropriately to match the research being conducted. The result is that there is no significant source of biosecurity information that can be used as a foundation for training and education. Therefore, the initial steps in this endeavor should focus on concerted efforts to collect relevant biosecurity practices and principles that can be adapted, scaled, and applied to research programs. Perhaps it is not too far-fetched to believe that agencies like the CDC, the Defense Threat Reduction Agency or others, would look to draft a guidance document in a similar fashion to the *BMBL* [15].

As more biosecurity guidance is developed and released, research institutions will gain experience implementing biosecurity programs. This developing content will continue to fortify training and education programs. Several publications, including this one, are contributing to the industry with thought-leaders and experienced professionals providing their expertise on the subject. In addition, several graduate-level programs in defense and security are now providing instruction on biosecurity policy. But what is likely to provide more awareness to biosecurity than any other medium is events as they are conveyed in the press. This was demonstrated repeatedly in the second half of 2014, with laboratory issues at the NIH and CDC [14]. Incidents such as these will stimulate the call for more policy and guidance on issues related to biosecurity, such as access control and inventory management.

As biosecurity guidance develops, as a priority and a product, training and education must continue. This will serve as the foundation for all actions that are taken by academic research institutions to fund, implement and oversee biosecurity principles and practices.

Summary and recommendations

This chapter has described the intersection of the roles and responsibilities of the IBC with essential elements of biosecurity. While many of the necessary practices are supplied by federal regulations and guidelines, many gaps remain. We propose several recommendations going forward:

- IBC members should be trained and educated in relevant biosecurity, such as access control, personnel reliability, threat assessments, etc. that would require additional training for the average practicing laboratory or clinical scientist.
- Institutional leadership should be engaged in and support the IBC's biosecurity-related activities, since it in the best interest of institutions carrying out biosecurity-related research to support their researchers in matters that could have national security implications.
- Government agencies (CDC, NIH, DoD, FBI) should work together to draft guidelines with a view toward minimizing additional administrative requirements on already resource-strapped research institutions.
- A "culture of biosecurity" should be developed at institutions with activities requiring security. This kind of institutional change would require early exposure to the concept of biosecurity, in keeping with calls for engendering a "culture of responsibility" among young scientists through Responsible Conduct of Research Training [5].

References

[1] Traxler EA, Tycko J, Wojno AP, Wilson JM. Lessons learned from the clinical development and market authorization of Glybera. Hum Gene Ther Clin Dev 2013;24(2):55–64.
[2] Imperiale MJ, Casadevall A. A new synthesis for dual use research of concern. PLoS Med 2015;12(4):e1001813.
[3] Biotechnology Research in an Age of Terrorism. 2004 NRC.
[4] 42 CFR 73.17; 7 CFR 331.17 and 9 CFR 121.17.
[5] Sutton V. Smarter regulations: comment on Responsible conduct by life scientists in an age of terrorism. Sci Eng Ethics 2009;15(3):303–9.
[6] Guidance for enhancing personnel reliability and strengthening the culture of responsibility: a report of the National Science Advisory Board for Biosecurity. Available from: <http://osp.od.nih.gov/sites/default/files/resources/CRWG_Report_final.pdf>.
[7] Beeching NJ, Fenech M, Houlihan CF. Ebola virus disease. Br Med J 2014;349:g7348.
[8] Inglesby TV, O'Toole T, Henderson DA, Bartlett JG, Ascher MS, Eitzen E, Working Group on Civilian Biodefense Anthrax as a biological weapon 2002; updated recommendations for management. J Am Med Assoc 2002;287(17):2236–52.
[9] WHO Biorisk Management: Laboratory Biosecurity Guidance 2006. Available from: <http://www.who.int/csr/resources/publications/biosafety/WHO_CDS_EPR_2006_6.pdf>
[10] Leitenberg M. Biological weapons in the twentieth century: a review and analysis. Crit Rev Microbiol 2001;27(4):267–320.
[11] Millett PD. The biological and toxins weapons convention. Rev Sci Tech 2006;25(1):35–52.
[12] Bakanidze L, Imnadze P, Perkins D. Biosafety and biosecurity as essential pillars of international health security and cross-cutting elements of biological nonproliferation. BMC Public Health 2010;10:S12.
[13] Reardon S. Safety lapses in US government labs spark debate. Nature/News 2014 July 16.
[14] Skorton D, Wynes DL, Marton G. Regulatory challenges in university research. Issues Sci Technol 2014;18(2).
[15] Wilson DE, Chosewood LC, editors. Biosafety in Microbiological and Biomedical Laboratories (BMBL) (5th ed.). U.S. Department of Health and Human Services; 2009. HHS Publication No. (CDC) 21-1112. Available from: http://www.cdc.gov/biosafety/publications/bmbl5/.

IBCs – A cornerstone of public trust in research

Kathryn L. Harris, Ryan Bayha, Jacqueline Corrigan-Curay and Carrie D. Wolinetz

Chapter Outline

Introduction

Recombinant DNA has been a transformative technology, providing tools that not only have enabled tremendous understanding of life at the most fundamental levels, but that have also led to a myriad of medical and agricultural applications. Progress in recombinant DNA research continues to revolutionize approaches to life science research and biotechnology and has been possible because scientists taking the lead in developing this technology had the foresight to recognize that the promise of

Ensuring National Biosecurity. DOI: http://dx.doi.org/10.1016/B978-0-12-801885-9.00012-3
© 2016 Elsevier Inc. All rights reserved.

recombinant DNA could only be realized if they assumed responsibility for addressing the safety and ethical concerns that it raised.

The current system of oversight of recombinant DNA research was established almost 40 years ago when the *NIH Guidelines for Research Involving Recombinant or Synthetic Nucleic Acid Molecules* (*NIH Guidelines*) were first written [1]. At the federal level, the *NIH Guidelines* were initially administered by the Office of Recombinant DNA Activities (ORDA) that later became the Office of Biotechnology Activities (OBA) within the Office of Science Policy.

The *NIH Guidelines* outline the requirements for local oversight, including the establishment of an institutional oversight committee. The first *NIH Guidelines* articulated the requirements for "Institutional Biohazards Committees" that were later renamed "Institutional Biosafety Committees" (IBCs) to more clearly reflect their role. IBCs must review recombinant and synthetic nucleic acid molecule research for conformity with the *NIH Guidelines*. In addition, they assess the research for potential risks to health and the environment. This is accomplished by reviewing physical and biological containment for the research and ensuring that researchers are adequately trained to conduct the research they are proposing safely.

The hallmarks of this oversight system from its inception were public participation and transparency. Attention to the concerns of the community and local interests is a major theme that carries forward in the system of biosafety oversight today. This key element has served to preserve public trust in the safety of the life sciences research enterprise. In retrospect, the risks of recombinant DNA technology that were feared early in its evolution did not materialize. That fact notwithstanding, the development of a scientifically based oversight system with the IBCs as the centerpiece permitted the safe development of recombinant DNA as an essential technology in research. Over the years, oversight by IBCs has proven critically important to ensuring safety throughout various research fields – medical, occupational, environmental – as well as in promoting responsible scientific practice. Due to the dynamic nature of the life sciences there remains an ongoing need to assess biosafety dimensions of the research being conducted and to manage any risks associated with work. As life sciences research continues to advance, many lines of research, particularly involving highly pathogenic organisms, continue to generate public concern. Financial support for life sciences research comes primarily from publicly derived tax dollars, and so the life sciences community must demonstrate to the public that it is being a responsible steward of those funds. IBCs today remain critically important in preserving public trust and thus facilitating continued scientific progress. The National Institutes of Health (NIH) and the institutions it funds must continue to ensure that IBCs are equipped to fulfill their responsibilities so that biosafety risks are responsibly managed and public safety and trust are preserved.

Evolution of an oversight framework – Asilomar and beyond

The landmark Asilomar Conference of 1975, involving leading scientists from a variety of disciplines, launched the development of what was to become an enduring system of institutional and federal oversight for research utilizing recombinant DNA technology guided by the NIH [2]. With the advent of recombinant DNA techniques in the early 1970s, a debate arose in the scientific community regarding the potential health and environmental risks of genetically manipulated organisms. The public also became involved in this debate. The concern about the dangers of emerging recombinant DNA technologies led to the scientific community calling for an unprecedented voluntary moratorium on certain experiments with recombinant DNA until the risks could be further characterized and procedures developed to minimize those risks. Scientists called for the formation of a national oversight body to ensure public discussion and ongoing oversight of this emerging technology [3]. One of the outcomes of this process was the formation of the National Institutes of Health Recombinant DNA Molecule Program Advisory Committee, subsequently renamed the Recombinant DNA Advisory Committee (RAC), a Federal advisory committee that was tasked with developing the first recombinant DNA research guidelines, formalized in the *NIH Guidelines for Research Involving Recombinant DNA Molecules (NIH Guidelines)*.

In the mid-1970s, recombinant DNA was a nascent and poorly understood technology, but was emerging as an important tool for conducting biomedical research. The public was apprehensive about this novel technology, related to the potential impact on the environment and public health, as well as a host of ethical and social issues due to its capacity to modify the molecular centerpiece of life. The debate over safety captured congressional attention, resulting in over a dozen legislative proposals that would have statutorily imposed a number of restrictions on the development and use of this technology [4]. A number of local jurisdictions were also considering similar ordinances that would have precluded the use of recombinant DNA altogether. Most notable among these was the city of Cambridge, Massachusetts, that in 1977 became the first jurisdiction in the United States to directly regulate basic scientific research using recombinant DNA [5]. The Cambridge Recombinant DNA Technology Ordinance also established strict oversight of university and commercial laboratories that engaged in recombinant DNA research. The requirements set forth in that city ordinance were based on the *NIH Guidelines* and the Cambridge Biosafety Committee carried out enforcement. The Committee, staffed by the Cambridge Public Health Department, is comprised of Cambridge residents and still operates to this day.

The *NIH Guidelines* as a dynamic oversight framework

The initial version of the *NIH Guidelines* [6] was drafted by the RAC a year after the Asilomar Conference. Unlike the statutes being contemplated, the *NIH Guidelines* were intended to be a "living" document detailing required biosafety practices for institutions and investigators to follow, consonant with current scientific understanding. By design the *NIH Guidelines* is a dynamic document that maintains its relevance by evolving along with science and technology and it has undergone multiple revisions since 1976. In 1978, almost immediately after the completion of the first iteration of the *NIH Guidelines*, the then-Department of Health, Education, and Welfare (now the Department of Health and Human Services) held a national conference to take a fresh look at the state of our understanding of risks posed by recombinant DNA technology. Up to that time, the RAC had reviewed every single recombinant DNA protocol that was conducted. NIH decided that RAC review of all protocols, as well as other restrictions on the field of research outlined in the first iteration of the *NIH Guidelines*, were no longer necessary. Instead, the transparency provisions of the oversight system and public access to the institutional review processes were enhanced, substituting for the diminished role of the RAC in protocol reviews. Toward that end, several revisions were made to the *NIH Guidelines* including significantly increased public access to information about recombinant DNA research activities and public participation in the administration of the *NIH Guidelines* in local communities. The *NIH Guidelines* originally required that no less than 20% of the IBC would be comprised of members who had no affiliation with the institution and were in a position to represent community interests. The changes to the *NIH Guidelines* eliminated the "no less than 20%" clause and instead, required there be at least two IBC members who had no affiliation with the institution and were in a position to represent community interests. Furthermore, important records of the IBC, including minutes, memoranda of understanding and agreement, registration documents, and other materials submitted to the Federal government had to be made available to the public upon request.

In the early 1980s, many observers of the field of recombinant DNA research were prescient enough to foresee its eventual application to humans. Among these were a coalition of religious leaders who asked then President Reagan's Commission for the Study of Ethical Problems in Biomedical and Behavioral Research – a group set up to look at the ethics of human subjects research – to undertake the special topic of genetic engineering with humans [7]. The result of that analysis was the report "Splicing Life: The Social and Ethical Issues of Genetic Engineering with Human Beings (1982)." This report was conveyed to the RAC for its consideration, which led the RAC to recommend a number of changes to the *NIH Guidelines* that went into effect in April 1984. These included giving IBCs the formal mandate to review and approve human gene transfer (HGT) research. Subsequently in 1986, a document titled "Points to Consider in the Design and Submission of Gene Therapy Trials to the NIH" was developed by the RAC, and incorporated into the *NIH Guidelines* as Appendix M. It outlined for IBCs the kinds of matters they were expected to evaluate when reviewing proposals for the administration of recombinant DNA to human subjects in clinical trials at a time of substantial advances in HGT, coincident with the first gene transfer trial being approved by the NIH Director in 1989.

In the mid-1990s, the United States Department of Agriculture (USDA) had a RAC-like body known as the Agricultural Biotechnology Research Advisory Committee (ABRAC), which was in the process of developing containment guidelines for agricultural research taking place outside the context of the laboratory – such as in greenhouses and large animal facilities. The NIH, in consultation with the Federal *ex-officio* members of the RAC, determined that it would make more sense to consolidate all of the requirements concerning the containment of recombinant DNA into the *NIH Guidelines*. Consequently, the USDA guidelines were incorporated into the *NIH Guidelines* as Appendices P (for plants) and Q (for animals). The addition of these appendices also included modifications to the IBC's membership requirements to ensure that appropriate expertise existed on the IBC when the institution was conducting this type of research.

The mid-1990s also proved to be a time of introspection for the NIH with regard to recombinant DNA oversight, especially in the realm of HGT research, which had become the leading matter brought before the RAC. The NIH director at the time called for a thorough analysis of the investment that NIH was making in the field of HGT research, and the optimal mechanisms for NIH oversight of the field. This was accomplished through the efforts of two committees: The Orkin-Motulsky Committee looked at the question of NIH's investment in the field and concluded that, while the field still had great promise, progress was hampered by gaps in scientific knowledge, and the Committee urged the NIH to fund proportionately more basic research to build a stronger scientific foundation that would inform clinical science [8]; the Verma Committee looked at optimizing the value of the RAC and concluded that it was not necessary for the RAC to publicly review every protocol submitted to the NIH, nor was it necessary for NIH to approve every protocol. IBCs and other bodies such as the Institutional Review Boards (IRBs) and the Food and Drug Administration (FDA) had approval authority, and the RAC process could be of optimal value as a means of informing the deliberations of these other groups. Consequently, they recommended that the RAC only review in public those protocols that raise special or novel scientific, safety, or ethical issues [9]. The recommendations of the Verma Committee were embodied in the October 1997 revisions to the *NIH Guidelines* and eliminated the need for the NIH to approve each HGT trial.

The *NIH Guidelines* have continued to be updated since 1997 to address various issues in both HGT and basic laboratory research. The last major change to the *NIH Guidelines* occurred in 2012 when they were amended to specifically cover certain basic and clinical research with nucleic acid molecules created solely by synthetic means. At the time, it was likely that most research with synthetic nucleic acids also involved the use of recombinant techniques and thus was already covered by the *NIH Guidelines*. However, this change was designed to be forward-looking, recognizing that new technologies can rapidly expand. The amendment to the *NIH Guidelines* also made it clear that the focus of biosafety review for new genetic constructs should focus on the product, not the technique used to produce it. Reflecting this change, the name of the *NIH Guidelines* was changed to the *NIH Guidelines for Research Involving Recombinant or Synthetic Nucleic Acid Molecules*.

The *NIH Guidelines* will never be complete as it is impossible to foresee all future advances in science. It is therefore important that all who are involved in both the conduct and the oversight of research adhere not only to the specifics of the *NIH Guidelines*, but also to the intent and spirit of the document. The dynamic nature of the *NIH Guidelines* has served to promote comprehensive oversight of a field that is continually advancing in a much more flexible manner than a statutory framework could, while permitting the science to proceed in a safe and responsible manner.

The NIH Guidelines today

Currently, the *NIH Guidelines* apply to research that is conducted at or sponsored by institutions receiving NIH funding for recombinant or synthetic nucleic acid molecule research. However, the *NIH Guidelines* have a "reach-through" provision. If an institution receives any NIH funding for projects involving recombinant or synthetic nucleic acid molecules, then all research involving recombinant or synthetic nucleic acid molecules conducted at or sponsored by the institution – even those that are not NIH-funded – become subject to the requirements of the *NIH Guidelines*. The logic for the broad applicability of the *NIH Guidelines* is that, to be effective, the same system of oversight has to be observed by all researchers at a given institution. Thus, the *NIH Guidelines* have become the universal standard for safe, scientific practice in recombinant and synthetic nucleic acid molecule research. As such, many private entities, federal agencies, and other institutions also follow the *NIH Guidelines* voluntarily even though they are not otherwise subject to its requirements.

The *NIH Guidelines* are termed "guidelines" because they establish principles and basic safety practices. The document articulates performance standards without being unduly prescriptive. The title of the document is not meant to convey, however, that the *NIH Guidelines* are optional. The requirement to abide by the *NIH Guidelines* is a term and condition of NIH funding for research that is subject to their scope. Noncompliance with the *NIH Guidelines* could result in suspension or termination of NIH funds for research with recombinant or synthetic nucleic acid molecules, or the requirement to have all the projects subject to the *NIH Guidelines* at the institution receive prior NIH approval. At the same time, the *NIH Guidelines* permit a great deal of flexibility, particularly in implementation of the administrative aspects of research oversight. This enables institutions to tailor their programs to meet specific needs. It is the responsibility of institutions to establish and implement policies for the safe conduct of research subject to the *NIH Guidelines*. A key part of this responsibility is the establishment of an IBC.

IBCs as the linchpin in our framework of biosafety oversight

Receipt of NIH funding for recombinant or synthetic nucleic acid molecule research comes with a requirement that the institution establish an IBC. The IBC must be composed of at least five individuals who collectively have the appropriate expertise to be able to review the research conducted at the institution (including plant or animal biosafety experts if applicable). In addition to their scientific expertise, members of the IBC should have knowledge of biosafety and physical containment principles, and have an understanding of the institution's safety policies. The institution must appoint a Biological Safety Officer (BSO), who serves on the IBC if the institution is conducting any high containment recombinant or synthetic nucleic acid research (i.e. at Biosafety Level 3 or 4) or any large-scale research (i.e. the use of cultures of greater than ten liters aggregate volume). At least two members of the IBC must not be affiliated with the institution and represent community interests. The requirement for two nonaffiliated members derives from the core principles of the NIH system of oversight – that IBC review and oversight be transparent and involve public participation.

Current scope of responsibilities

Under the *NIH Guidelines*, IBCs are responsible for local oversight of recombinant and synthetic nucleic acid molecule research (although their responsibilities need not be restricted to such research), and review proposed experiments to ensure that they are conducted in a manner consistent with the biosafety practices outlined in the *NIH Guidelines*. With respect to HGT trials, the IBCs have additional responsibilities (under Appendix M of the *NIH Guidelines*) for reviewing these protocols to ensure the safe and proper design of this research, including the analysis of adverse event reports and findings from animal studies germane to the design and conduct of human studies. Accordingly, specific IBC responsibilities articulated in the *NIH Guidelines* include:

- Reviewing research conducted at or sponsored by the institution for compliance with the *NIH Guidelines*. Such review requires different levels of evaluation depending on the nature of the work.
- Helping investigators determine the appropriate containment conditions in which to conduct their research. IBC recommendations are guided by one of several appendices in the *NIH Guidelines* that specify safety and containment practices for the various forms of research subject to the document.
- Assessing the adequacy of facilities, institutional procedures and practices, and investigator training and expertise for the type(s) of research being conducted.
- Periodically reviewing research conducted at the institution to ensure ongoing compliance.
- Adopting emergency plans covering accidental spills and personnel contamination resulting from research activities.
- Reporting any significant problems with or violations of regulations and any significant research-related accidents or illnesses to the appropriate institutional official and relevant federal entities.
- Ensuring investigators and laboratory staff are adequately trained to conduct all proposed work safely.

Importance of partnerships for promoting IBC excellence

It is incumbent upon both federal funders of research and the institutions that conduct the work to ensure that IBCs are well positioned to serve their pivotal role in upholding biosafety standards and practices, thereby sustaining public trust in the life sciences research enterprise. At the institutional level, this means having robust, well-run biological safety programs. The heart of these programs is the IBC. Supporting IBCs and promoting the importance of biosafety from the level of individual researchers all the way to senior levels of institutional administration is a shared responsibility of the NIH and the research community who must work together toward ensuring the optimal functioning of IBCs.

NIH's role in supporting IBCs

The NIH has made it a priority to develop a program of outreach and education that is multifaceted and uses multiple modalities to communicate with IBCs and other elements of the institutional infrastructure with key responsibilities in upholding the biosafety principles and practices of the *NIH Guidelines*. NIH's outreach program is extensive and includes giving presentations and briefings at major national and regional scientific conferences, and professional development meetings of key stakeholder groups on compliance with the *NIH Guidelines*. Outreach staff develop and conduct individualized workshops and training sessions to educate about the NIH system of oversight of recombinant and synthetic nucleic acid molecule research under the *NIH Guidelines*. Target audiences include investigators, IBC members and staff, BSOs, industry representatives, and institutional compliance officials. Of particular note are the NIH-developed and -supported national training courses: "IBC Basics" and "Effective IBCs" that provide in-depth information on establishing and running an effective and compliant IBC program. In addition, NIH also hosts the IBC professional development conference series, a biannual event that is the only national professional development forum for members and staff of IBCs. NIH has also developed an array of electronic and printed educational materials including training presentations which institutions can tailor to their own needs, sets of Frequently Asked Questions, biosafety guidance tools on many topics of interest to institutions subject to the *NIH Guidelines*, brochures on investigator responsibilities, and posters to promote awareness of biosafety and biosecurity policies.

The NIH IBC site visit program

In November 2006, NIH launched the IBC site visit program [10]. The program is an essential element of NIH's outreach activities since it offers a tailored and interactive experience, providing a forum for institutions to ask questions about the *NIH Guidelines* and enabling them to make informed enhancements to their IBC programs to incorporate best practices and remedy any deficiencies. To date, NIH has visited a diverse set of over 100 institutions including universities, medical schools, research institutes, and commercial entities that conduct varied types of research programs, with a range of emphasis from basic biomedical or agricultural programs to clinical research. In addition to helping institutions improve their programs in the immediate term, this program aims to assist the IBC community in the longer term by:

- identifying common challenges facing institutions for the purpose of customizing NIH educational programs to assist institutions in overcoming them;

- developing a body of best practices that institutions may consider, as appropriate, to optimize the functioning of their own programs; and
- creating a self-evaluation tool that institutions can use to assess and improve their IBC programs.

Lessons learned from the site visit program

One of the major goals of the site visit program was to develop a body of information that could be shared with the IBC community with an eye towards enhancing IBC function. A number of common features and practices have been observed at institutions that have particularly well-run and effective programs:

- **Charter and standard operating procedures (SOPs)**
  An IBC charter and well-written SOPs can foster the development of a consistently well-run and compliant biosafety program. These documents define the authority of the IBC, outline its operating and review practices, describe protocol registration requirements that investigators must follow, and keep IBC members and investigators alike informed of the biosafety standards and review requirements that must be observed at the institution.
- **Mechanisms for ongoing review and oversight of research**
  The IBC is not simply responsible for approving research, but also for ensuring the ongoing safe conduct of that research. Under the *NIH Guidelines*, the IBC must conduct "periodic" review of research involving recombinant DNA. The *NIH Guidelines* do not prescribe the frequency of review, because the nature and biosafety profile of research involving recombinant DNA is extremely diverse, and the IBC must make its own judgments about the appropriate frequency with which to revisit ongoing work.
- **Robust training programs for IBC members, investigators and laboratory staff**
  The *NIH Guidelines* place a great deal of emphasis on training and education. Institutions are responsible for ensuring the training of everyone involved in the conduct or oversight of recombinant DNA research and conducting training is also a responsibility of BSOs and PIs. A clear understanding of the biosafety practices and individual responsibilities articulated in the *NIH Guidelines* is pivotal to an effective biosafety program.
- **Records of IBC activities**
  Committee minutes are critically important for a number of reasons. First, they create institutional memory and create a record of Committee decision-making that, in turn, fosters consistency in research reviews. Second, they document the IBC's fulfillment of its responsibilities under the *NIH Guidelines*. Finally, as part of the transparency fostered by the *NIH Guidelines*, IBCs must make their meeting minutes available to the public upon request, providing an important view into institutional safety practices.

Impact of the site visit program

Since its inception, there is strong evidence that the NIH site visit program has enhanced the effectiveness of institutional biosafety programs. Hackney et al. [11] presented the results of a 2010 survey conducted to examine trends in IBC practices. The authors compared their results to similar surveys that they conducted in 2002 and 2007. The results of the 2010 survey showed that IBCs have made significant improvements since 2002 in levels of staffing support, conducing training, and increasing transparency to the public and overall compliance with the *NIH Guidelines*. A primary reason cited by the authors for the improvement in IBC performance is the NIH site visit program. Of the institutions that responded to the survey, 80% stated that they believed the NIH site visit enhanced the institutions' oversight of recombinant and synthetic nucleic acid molecule research. The results of the survey also showed that institutions that have undergone site visits were better poised to effectively carry out their responsibilities under the *NIH Guidelines* than institutions that have not had a site visit. This can be attributed to the awareness-raising nature of the NIH site visit program. For example, the survey found that 85% of institutions that had a site visit have reported an incident involving a significant problem, violation of the *NIH Guidelines*, or a significant research-related accident or illness to NIH, while only 25% of the non-site-visited institutions had reported such incidents as required by the *NIH Guidelines*. Similarly, 70% of respondents who had received a site visit have PI training in place, whereas only 44% of institutions that have not had a site visit have similar training. The NIH has noted that incident reporting compliance increases after a site visit because the importance of incident reporting is strongly emphasized during the site visits.

Enhanced institutional compliance with the requirements of the *NIH Guidelines* may also be attributed to institutions utilizing the *Institutional Biosafety Committee Self-Assessment Tool* [12]. NIH first issued the Self-Assessment Tool in 2009, and revised the tool in 2014. The aim of the tool was twofold. First, it was quickly realized that NIH would not be able to perform a site visit to all of the registered IBCs (~890 in 2014). The Self-Assessment Tool was created to encourage institutions to assess their own programs using the same performance indicators as NIH does during a visit. The tool consists of 83 core questions relating to specific requirements of the *NIH Guidelines*. The tool poses a question, lists the citation of the *NIH Guidelines* that is pertinent and then provides NIH's guidance on how the requirement should be addressed by the institution. The tool also contains NIH's recommendations on a number of best practices not specifically articulated as requirements in the *NIH Guidelines* but nonetheless could enhance the effectiveness of the IBC.

Institutional roles in supporting IBCs

The *NIH Guidelines* articulate a number of responsibilities for institutions conducting research subject to the scope of the document, including the establishment and implementation of policies that provide for the safe conduct of the research. To effectively carry out their functions, it is essential that IBCs receive the full support and collaboration of senior institutional officials. IBCs must have the authority to fulfill their responsibilities properly. IBCs not only approve research, but also may reject proposed activities or shut down ongoing research if there are concerns about an activity meeting the biosafety standards articulated in the *NIH Guidelines*. IBCs must have the backing of senior institutional administration if they are to exert this authority meaningfully.

IBCs need to be sufficiently staffed and resourced to properly fulfill their responsibilities. One benchmark is to consider the number of protocols reviewed annually by the IRB, Institutional Animal Care and Use Committee (IACUC) and IBC, and consider whether the resources accorded each committee are in proportion to their respective workloads. In addition to the IBC's protocol review and approval responsibilities, IBCs and biosafety program staff should conduct training, laboratory inspections, and ongoing oversight of research. As research portfolios expand, institutions should also periodically conduct thorough assessments of the resources necessary for their programs to effectively fulfill the roles and responsibilities articulated in the *NIH Guidelines*, taking into account all of the responsibilities of the institution, the IBC and the biosafety staff. The NIH will continue to provide outreach and education resources to assist institutions, but encourages all institutions to take a rigorous and ongoing examination of their IBC programs.

The future face of IBCs

The changing research landscape and emerging life science technologies

Although many of the fears about the impact of recombinant DNA technology did not come to pass, many aspects of life sciences research continue to warrant review and oversight. The pace at which scientists can bring about changes in biological systems is ever-increasing. Experiments that took weeks or months to complete only a few years ago can now be performed in days or even hours. At the same time, there is increasingly more interdisciplinary research being conducted, with life science technologies being employed in areas such as material science and chemical engineering. Synthetic biology is a prime example combining methods, principles and knowledge from disciplines including biology, engineering, mathematics and computational science. Thus, IBCs will almost certainly be faced with reviewing increasingly more novel protocols in the interdisciplinary realm.

In addition to conducting risk assessments, an important responsibility of IBCs is to ensure that those conducting the research have the experience and training to conduct the research safely. It is likely that some scientists educated in fields other than traditional life sciences disciplines may not have had comprehensive biosafety training and may be unfamiliar with potential risks and how best to manage them. In such instances it will be particularly important that IBCs ensure that the risks of the research are evaluated as rigorously as possible and that all researchers involved have been appropriately trained.

Genome editing technologies

Scientists have recently developed several new genetic engineering tools, including genome editing technologies such as CRISPR [13], TALENS and zinc finger nucleases. Such new tools and capabilities may transform biological research, further benefiting human heath, agriculture, industry and the environment. While many of these new technologies may pose minimal biosafety risk, emerging technologies can raise uncertainties about their environmental safety and effects on human health and it is incumbent on those conducting and overseeing such research to ensure that potential risks are evaluated and managed as necessary. Data-driven risk assessments and commensurate oversight are key to demonstrating that the life science community continues to uphold a culture of responsibility in ensuring that science proceeds safely.

HGT research

In 2012, the NIH was asked by the American Society for Gene and Cell Therapy, the professional organization of scientists who conduct HGT research, to review the role of the RAC to determine whether in-depth individual reviews of every HGT protocol were still warranted given the state of the science after many years of conducting such trials, and whether the scientific, safety, ethical, and other concerns continue to justify a special level of oversight for this area of research. A review was then carried out as part of NIH's ongoing evaluation of its oversight framework for gene transfer and recombinant DNA research. In 2013, the NIH Director commissioned the Institute of Medicine (IOM) to establish an independent committee to specifically look at the question of whether HGT research raises issues of concern that warrant extra oversight by the RAC in the form of the review of individual clinical protocols and, that if such oversight was warranted, to recommend criteria to guide when the RAC should review this research.

In the report "Oversight and Review of Clinical Gene Transfer Protocols: Assessing the Role of the Recombinant DNA Advisory Committee" [14], the IOM committee concluded that although gene transfer research continues to raise important scientific, social and ethical questions, not all gene transfer research is novel enough or controversial enough to justify the current forms of additional oversight at the national level. The committee also recommended that all individual protocols should continue to be registered with the NIH, but these protocols should not be subject to public review by the RAC except in exceptional circumstances. More specifically, the IOM recommended that individual gene transfer protocol reviews by the RAC be restricted to situations where protocol review could not be adequately performed by other regulatory and oversight processes (for example, IBCs and IRBs). The IOM report stated that all protocols should continue to be registered with NIH and that review of adverse events on protocols should continue as the evaluation of trends may lead to greater awareness of safety concerns.

The NIH carefully considered and accepted the IOM recommendations on RAC review of HGT research [15]. If the IOM recommendations are implemented, protocols will continue to be registered with the NIH, but in-depth, public RAC review of individual gene transfer protocols will be limited to exceptional cases, such as when IBCs and other oversight bodies may need assistance in reviewing exceptionally novel protocols. The NIH will likely continue to provide key information on protocols, including the evolution of their design and details on the products used, to augment institutional review bodies' resources.

Research on highly pathogenic agents

Infectious diseases are a significant cause of mortality and morbidity. Diarrheal diseases, malaria, tuberculosis and influenza exact a huge toll on human health [16]. In addition, exotic pathogens associated with high morbidity and mortality in humans frequently emerge – a recent example being the highly lethal Middle East respiratory syndrome coronavirus that emerged in 2012. At the NIH, the National Institute of Allergy and Infectious Diseases (NIAID) research portfolio has expanded considerably in recent years in response to new challenges such as biodefense and emerging and re-emerging infectious diseases. While research on pathogenic diseases is vitally important to the design of appropriate medical countermeasures against them in the form of diagnostics, therapies and vaccines, such research, particularly work conducted in high and maximum containment laboratories, has been frequently scrutinized in terms of the potential risks posed to the public from laboratory accidents or intentional misuse [17].

The *NIH Guidelines* require that IBCs are established specifically for the review of research involving recombinant and synthetic nucleic acid molecules, and IBCs have the responsibility to ensure that such research activities are performed with appropriate biosafety precautions. However, because of their expertise, many institutions have assigned these committees additional authority, which may include the oversight of research involving other biohazardous materials, such as non-recombinant infectious agents. Other guidance documents, such as the CDC/NIH publication *Biosafety in Microbiological and Biomedical Laboratories (BMBL)* [18], suggest that IBCs should have a broader purview to determine the appropriate biosafety levels for experiments with pathogenic agents. Surveys of IBCs [11] indicate that the vast majority of IBCs do in fact have a broader purview than that assigned to them specifically under the *NIH Guidelines*. Many academic institutions are investing in high-containment laboratory facilities and as research portfolios expand into the arena of highly pathogenic diseases it will become ever more important that IBCs are poised to ensure the research is conducted as safely as possible with well-trained research staff using robust standard operating procedures in well-maintained facilities.

Oversight of dual use research of concern

Life sciences research is vital to improving public health, agriculture and the environment, and to strengthening our national security and economy. Yet the very research designed to find ways to better the health, welfare and safety of humankind can also yield information or technologies that could potentially be misused for harmful purposes. For instance, information from certain life sciences research could be misapplied to weaponize dangerous pathogens, to bypass or diminish the effectiveness of medical countermeasures, or to threaten in other ways the health and safety of humans, animals, plants and the environment.

Research yielding new technologies or information with the potential for both benevolent and malevolent applications is referred to as "dual use research." The dual use potential of certain life sciences research has been recognized as an important biosecurity issue for a number of years. Some degree of dual use potential may be inherent in a significant portion of life sciences research. However, the small subset of life sciences research with the highest potential for yielding knowledge, products or technology that could be misapplied to threaten public health or national security is referred to as "dual use research of concern." The US Government (USG) has defined "dual use research of concern" as:

> *Life Sciences research that, based on current understanding, can be reasonably anticipated to provide knowledge, information, products, or technologies that could be directly misapplied to pose a significant threat with broad potential consequences to public health and safety, agricultural crops and other plants, animals, the environment, materiel, or national security.*

It is vitally important that researchers and their institutions are vigilant with respect to the dual use potential of life sciences research that they carry out.

In September 2014, the USG issued a *Policy for Institutional Oversight of Life Sciences Dual Use Research of Concern* [19]. The policy requires research institutions to establish a process for identifying dual use research of concern and articulates and formalizes the roles and responsibilities of institutions and investigators when they are conducting research supported by the Federal government that falls under the scope of the policy, with the aim of preserving the benefits of life sciences research while minimizing the risk that the knowledge, information, products or technologies generated by such research could be used in a manner that results in harm. Institutions subject to the policy are required to establish and implement internal policies and practices that provide for the identification and effective oversight of dual use research of concern. A key responsibility is the establishment of an Institutional Review Entity (IRE) to execute the requirements of the policy. While the policy states that "a range of mechanisms for fulfilling the role of an IRE are acceptable as long as the review entity is appropriately constituted and authorized by the institution to conduct the dual use review," it is likely, that in many instances, institutions will choose to utilize their IBCs to perform the functions of the IRE. The NIH and the other Federal government agencies that conduct or support life sciences research, have developed a number of resources to assist IREs in their review of research that is potentially dual use research of concern.

Conclusion

In 2015, 40 years after Asilomar, IBCs continue to play a relevant and vital role in the life sciences oversight system and have proven to be a key component in the biosafety management of research activities. Today, our understanding of recombinant DNA technology and its attendant risks is far greater, and although many of the initial fears about the technology have turned out to be unfounded, there remains the need to be vigilant over many aspects of the research enterprise. Capabilities in the life sciences have advanced rapidly in the past four decades and our capacity to manipulate organisms is likely to continue to advance at a similar, if not accelerating, pace for the foreseeable future.

The scientific community has long demonstrated a culture of responsibility in ensuring the safety of researchers, the public and the environment, but must also remain cognizant of the potential risks of emerging technologies and manage them accordingly. The public continues to be pointedly concerned about many aspects of life sciences research, especially in the context of human studies and research involving infectious agents. While the public continues to support biomedical research, its trust in the endeavor may be fragile. The IBCs have been central to a transparent demonstration of the ability of the scientific community to serve as responsible stewards of publicly funded research, and in helping to earn and preserve the trust placed in it.

Acknowledgments

The authors wish to thank Allan C. Shipp for his helpful comments on this chapter.

References

[1] NIH Guidelines for Research Involving Recombinant or Synthetic Nucleic Acid Molecules. 2013.

[2] Berg P, Baltimore D, Brenner S, Roblin R, Singer MF. Asilomar conference on recombinant DNA molecules. Science 1975;188(4192):991–4.

[3] Berg P, Baltimore D, Boyer HW, Cohen SN, Davis RW, Hogness DS, et al. Potential hazards of recombinant DNA molecules. Science 1974;185(4148):303.

[4] Talbot B. Development of the National Institutes of Health Guidelines for recombinant DNA research. Public Health Rep 1983;98(4):361–8.

[5] Lipson S. The cambridge model. Gene Watch 2003;15(5):7–10.

[6] Recombinant DNA Research-Guidelines. Fed Reg 1976;41 (131, pt. II): 27902–27943.

[7] President's commission for the study of ethical problems in medicine and biomedical and behavioral research. Splicing Life: a report on the social and ethical issues in genetic engineering with human beings. 1982.

[8] Orkin SH, Motulsky AG. Report and recommendations of the panel to assess the NIH Investment in research on gene therapy. 1995.

[9] Ad Hoc Review Committee-Recombinant DNA Advisory Committee. Verma Report: Executive summary of findings and recommendations. National Institutes of Health; 1995. Available from: <http://osp.od.nih.gov/office-biotechnology-activities/verma-report>.

[10] Bayha R, Harris KL, Shipp AC, Corrigan-Curay J, Wolinetz CD. The NIH office of biotechnology activities site visit program: Observations about institutional oversight of recombinant and synthetic nucleic acid molecule research. Application of Biosafety 2015;20(2):75–80.

[11] Hackney RW, Myatt TA, Gilbert KM, Caruso RR, Simon SL. Current trends in Institutional biosafety committee practices. Application of Biosafety 2011;17(1):11–14.

[12] Institutional Biosafety Committee Self-Assessment Tool. National Institutes of Health; 2014. Available from: <http://osp.od.nih.gov/office-biotechnology-activities/biosafety-guidance-institutional-biosafety-committees/ibc-self-assessment-2014>.

[13] Pennisi E. The CRISPR craze. Science 2013;341(6148):833–6. Available from: http:// www.sciencemag.org/content/341/6148/833.summary%3Fintcmp%3Dcollection-crispr.

[14] Lenzi RN, Altevogt BM, Gostin LO, editors. Oversight and review of clinical gene transfer protocols: assessing the role of the recombinant DNA advisory committee. Washington, DC: National Academies Press; 2014;374–8.

[15] Collins FS. Statement by the NIH Director on the IOM report addressing the role of the Recombinant DNA Advisory Committee in oversight of clinical gene transfer protocols. National Institutes of Health; 2014. Available from: http://www.nih.gov/about/ director/05222014_statement_iom_rac.htm.

[16] Fauci AS. New and reemerging diseases: the importance of biomedical research. Emerging Infectious Diseases 1998;4(3):374–8. Available from: http://wwwnc.cdc.gov/ eid/article/4/3/98-0308.

[17] Gottron F, Shea DA. Oversight of high-containment biological laboratories: Issues for Congress. Washington, DC, USA: Congressional Research Service; 2009.

[18] Wilson DE, Chosewood LC, editors. Biosafety in Microbiological and Biomedical Laboratories (BMBL) (5th ed.). U.S. Department of Health and Human Services; 2009. HHS Publication No. (CDC) 21-1112. Available from: http://www.cdc.gov/biosafety/ publications/bmbl5/.

[19] United States Government Policy for Institutional Oversight of Life Sciences Dual Use Research of Concern. U.S. Government Science, Safety, Security (S3); 2014. Available at: <http://www.phe.gov/s3/dualuse/Pages/default.aspx>.

Strengthening the role of the IBC in the 21st century

R. Mark Buller, Nancy D. Connell, Steven S. Morse, Mark Campbell
and Raymond C. Tait

Chapter Outline

Introduction

Over the last 30 years the recombinant DNA revolution has catalyzed our discovery of fundamental principles in biology, which continue to lead to translational applications in human medicine and agriculture. This new era started in 1975 when a small group of scientists, recognizing the potential for recombinant DNA technology to work for and against humankind, met at Asilomar to discuss a way forward [1]. A self-imposed moratorium on recombinant DNA research gave way to experimentation following the principles of the Asilomar Conference, and codes of practice formulated by national bodies in various countries with significant input from scientists [2]. This resulted in the development of a transparent, best-practices system of science that assured the potential of recombinant DNA research could be achieved in a safe manner, and remains a model for regulation of science by scientists [3]. In the United States, the first iteration of the Recombinant DNA Research

Ensuring National Biosecurity. DOI: http://dx.doi.org/10.1016/B978-0-12-801885-9.00013-5

Guidelines (National Institutes of Health (NIH) Guidelines) were drafted with public input in 1976 by a National Institutes of Health federal advisory committee that eventually became known as the Recombinant DNA Advisory Committee (RAC) (see also Chapter 12) [4]. Initially the RAC reviewed all Principal Investigator (PI)-initiated proposals (referred to as registration documents) that involved recombinant DNA, but since 1978 the review and approval of most registration documents has been conditionally delegated to the Institutional Biosafety Committee (IBC) at research establishments (see also Chapters 2 and 12). The drafted NIH *Guidelines* were not static and were modified as new technologies were developed, and as our understanding of risk evolved from intuitive to empirical based on new knowledge. Led by CDC and NIH initiatives, the NIH *Guidelines* were the starting point in a partnership with the life sciences community to develop a code of practices for biosafety in microbiological and biomedical laboratories. In 1984 this collaboration resulted in the publication of *Biosafety in Microbiological and Biomedical Laboratories (BMBL)* [5].

For the last 40 years best practices based on the NIH *Guidelines* and the *BMBL* have helped ensure the safety of the laboratory worker and the public. The rise of nonstate-sponsored terrorism in the latter part of the last century, the recognition that biotechnology was pervasive, and that life sciences research had a dual-use nature, required new approaches to identify and evaluate the research risk. In response to this and other threats, the government enacted the Antiterrorism and Effective Death Penalty Act of 1996, the USA PATRIOT Act of 2001, and Bioterrorism Preparedness and Response Act of 2002 (see also Chapters 1 and 2). Together this legislation made it illegal to possess biological agents or toxins for nonpeaceful purposes. With regard to the most dangerous pathogens and toxins known as biological select agents and toxins (BSAT), these acts mandated their registration, criminal background checks for users, a regulatory system for storage and transportation, and the creation of a class of "restricted persons," who were prohibited from their access (see also Chapter 10). Because implementation of this legislation could have the unintended consequence of making the United States more vulnerable to bioterrorism by slowing or blocking the development of the very diagnostics, prophylactics and therapeutics that are needed to protect the country from bioterrorism, the National Research Council of the National Academies of Sciences formed the Committee on Research Standards and Practices to Prevent the Destructive Applications of Biotechnology. The charge of this committee was:

> ...to prevent the destructive application of biotechnology research and to recommend changes in these practices that could improve U.S. capacity to prevent the destructive application of biotechnology research while still enabling legitimate research to be conducted. Ref. [6]

The Committee focused on biotechnology research with the potential to cause catastrophic harm through the misuse of BSAT, as well as research that could:

> ...facilitate the creation of novel pathogens with unique properties or create entirely new classes of threat agents. Ref. [6]

In 2004 the Committee published its report entitled "Biotechnology Research in an Age of Terrorism" (NRC 2004 Report), which provided seven recommendations (Table 13.1) that together were designed and intended to minimize impediments to fundamental research, while identifying the research with the greatest potential for misuse for additional scrutiny.

The NRC 2004 Report recommendations concerning dual-use research (DUR) with the most dangerous pathogens and toxins would be implemented by self-governance of the scientific community and existing regulatory practice. Of its seven recommendations, recommendations 1, 2, 4, and 5 were pertinent to the IBC. Recommendation 1, arguably one of the most important, proposed comprehensive education of persons involved in life sciences research with BSAT with regard to the dual-use dilemma. The remaining three recommendations focused on the actual system of regulation of DUR with BSAT. Recommendation 2 suggested that the IBC, which already reviewed experiments involving recombinant and synthetic DNA and infectious agents, would also review DUR with BSAT that belonged to seven categories (see below). Recommendation 4 proposed the creation of a National Science Advisory Board for Biosecurity (NSABB). The NSABB was proposed to provide case-specific oversight of research and its communication and dissemination, if relevant to national security and biodefense (see also Chapter 6). In addition, it would act in an advisory capacity to alert the government to novel findings that have national security implications. Furthermore, the NSABB would serve as a resource providing education outreach [7,8], advice to journals concerning DUR (e.g., 2011 influenza A/H5N1 gain-of-function controversy), and international engagement with scientists and professional organizations [9]. Furthermore, the NSABB would conduct periodic review of the implementation of current legislation and existing regulation to ensure an optimal balance is maintained between stimulating life-sciences research, and ensuring national security (recommendation 5). And finally, the NRC 2004 Report recognized the key role federal funding would play in the development of this new system:

> ...successfully implementing the system ...will require significant additional resources at each stage; we do not attempt to provide an estimate of those costs. Otherwise, concerns for unfunded mandates could be a significant barrier to full consideration of the proposal by the scientific community. Ref. [6]

In 2007 the NSABB published a seminal report entitled: "Proposed Framework for the Oversight of Dual-Use Life Sciences Research: Strategies for Minimizing the Potential Misuse of Research Information" (NSABB 2007 Report) [10].

This report continued the evolution of a system for the oversight of DUR that was begun with the NRC 2004 Report, but with an important distinction. The NSABB 2007 Report recognized that DUR could be generated in the life sciences and aligned fields with a myriad of various possibilities, not just in the context of BSAT. The report coined the phrase dual-use research of concern (DURC) to describe a subset of life sciences research that could be directly misused to cause harm. This report proposed:

> ...a framework for the development – by the federal government – of a comprehensive system for the responsible identification, review, conduct, and communication of dual-use research. Ref. [10]

Table 13.1 **Recommendations from biotechnology research in an age of terrorism**

Recommendation	Description	Action
1	Educating the Scientific Community	We recommend that national and international professional societies and related organizations and institutions create programs to educate scientists about the nature of the dual use dilemma in biotechnology and their responsibilities to mitigate its risks.
2	Review of Plans for Experiments	We recommend that the Department of Health and Human Services (DHHS) augment the already established system for review of experiments involving recombinant DNA conducted by the National Institutes of Health to create a review system for seven classes of experiments (the Experiments of Concern) involving microbial agents that raise concern about their potential for misuse.
3	Review at the Publication Stage	We recommend relying on self-governance by scientists and scientific journals to review publications for their potential national security risks.
4	Creation of a National Science Advisory Board for Biosecurity	We recommend that the Department of Health and Human Services create a National Science Advisory Board for Biosecurity (NSABB) to provide advice, guidance, and leadership for the system of review and oversight we are proposing.
5	Additional Elements for Protection Against Misuse	We recommend that the federal government rely on the implementation of current legislation and regulation, with periodic review by the NSABB, to provide protection of biological materials and supervision of personnel working with these materials.
6	A Role for the Life Sciences in Efforts to Prevent Bioterrorism and Biowarfare	We recommend that the national security and law enforcement communities develop new channels of sustained communication with the life sciences community about how to mitigate the risks of bioterrorism.
7	Harmonized International Oversight	We recommend that the international policymaking and scientific communities create an International Forum on Biosecurity to develop and promote harmonized national, regional, and international measures that will provide a counterpart to the system we recommend for the United States.

In addition to proposing a detailed process for evaluation of DUR at the institutional level through an Institutional Review Entity (IRE), the report emphasized the importance of mandatory education and training of all life scientists at federally funded institutions, a code of conduct for scientists and laboratory personnel working in the life sciences and aligned fields, and public involvement in the DUR debate. Importantly, the NSABB 2007 Report reiterated a concern of the NRC 2004 Report that oversight of DUR should not become another unfunded mandate.

Eleven years after the NRC 2004 Report and 8 years after the NSABB 2007 Report, the US government (USG) has not fully implemented a robust system to evaluate DUR. In particular, the USG has failed to effectively fund, support, and expand the role of the local Institutional Biosafety Committee in the oversight of DUR and in the changing scientific landscape; a comprehensive education program for all in life sciences research with regard to the dual-use dilemma has not been forthcoming; and finally, there has been no systematic evaluation of the impact of USG policy, regulations, and guidance on an institution's cost structure and on scientific discovery. Current USG policy appears to diminish the role of the IBC in the oversight of DURC, whereas we believe the local IBCs should assume yet even more responsibility. We detail our judgments on current USG DUR policy and provide recommendations for future oversight of DUR from our perspective as senior administrators and laboratory scientists charged with the responsibility of conducting life-sciences research in an era of increasing regulatory requirements and decreasing federal support.

IBC review of DURC

US government (USG) policy and guidelines on DURC

The USG DURC Policy (2015 DURC Policy) was finalized as of September 24, 2014 with an effective start date of September 24, 2015. The 2015 DURC Policy definition is a modification of the NSABB 2007 Report DURC definition and reads as:

> *Research that, based on current understanding, can be reasonably anticipated to provide knowledge, information, products, or technologies that could be directly misapplied to pose a significant threat with broad potential consequences to Public Health and safety, agriculture crops and other plants, animals, the environment, materiel, or national security. Ref. [11]*

The scope of these USG policies is limited to life sciences research involving tier 1 BSAT and reconstructed 1918 influenza virus and avian influenza virus (15 BSAT) *and* seven categories of experiments (Table 13.2). Because research with the 15 BSAT represents a small fraction of life sciences research, the current USG policies do not provide oversight for most DUR in the life sciences or for training and education of the involved scientists as called for in the NSABB 2007 Report. Instead, it provides additional layers of scrutiny to a small segment of research in BSL-3 and BSL-4 laboratories, and targets its personnel for additional training and education. An even

Table 13.2 **Dual use of research of concern**

Category	Dual use of research of concern	Relevant section of NIH guidelines
1	Enhances the harmful consequences of the agent or toxin	Section III-D-3
2	Disrupts immunity or the effectiveness of an immunization against the agent or toxin without clinical or agricultural justification	
3	Confers to the agent or toxin resistance to clinically or agriculturally useful prophylactic or therapeutic interventions against that agent or toxin or facilitates their ability to evade detection methodologies	Section III-A-1-a, Section III-D-7-d
4	Increases the stability, transmissibility, or the ability to disseminate the agent or toxin	Section III-D-4
5	Alters the host range or tropism of the agent or toxin	Section III-D-3
6	Enhances the susceptibility of a host population to the agent or toxin	Section III-D-4, Section III-D-3
7	Generates or reconstitutes an eradicated or extinct agent or toxin listed in Section 6.2.1 of 2015 DURC policy [11]	Section III-B-1, Section III-D-7-c

smaller segment of research will receive additional review at the US Department of Health and Human Services (DHHS) prior to a funding decision based on: "A framework for Guiding US Department of Health and Human Services (DHHS) Funding decisions about Research Proposals with the Potential for Generating Highly Pathogenic Avian Influenza H5N1 Viruses that are Transmissible among Mammals by Respiratory Droplets" (2013 DHHS framework) [12]. Although the 2015 DURC Policy and the 2013 DHHS framework thoroughly scrutinize a small segment of life sciences research for DURC, the authors of the 2015 DURC Policy recognized that DUR exists outside of the current scope and stated:

> *Institutions have discretion to consider other categories of research for DURC potential. Ref. [11]*

However, it is unlikely that a majority of institutions will undertake an expanded review of DUR without the publication of additional USG policies. For a thorough discussion of 2015 DURC Policy, see Chapter 6.

DURC

The USG policies with regard to DURC are far too limited. In addition to BSAT research in high-containment BSL-3/BSL-4 laboratories, many other potential sources of DURC may arise from scientists working in low-containment BSL-1 or BSL-2 laboratories or amateur biologists experimenting in their garages. A well-publicized example of DUR that can be considered DURC came from a project that used ectromelia virus to develop an immune-contraceptive vaccine for wild mice. Ectromelia virus is a risk group 1 agent that is 95% identical to variola virus at the nucleotide level and is used to model smallpox infections. Ectromelia virus was modified by recombinant DNA techniques to express a gene for murine zona pellucida glycoprotein-3 (ZP-GP-3), which is part of an extracellular matrix surrounding the developing mammalian oocyte. Upon infection of mice with this ectromelia-ZP-GP-3 recombinant virus, the immune system of the mouse recognized the ZP-GP-3 as a foreign antigen, and synthesized antibodies against it. These anti-ZP-GP-3 antibodies subsequently attacked the developing oocyte and ovaries of the mouse resulting in sterility of about 6 months duration. This is an example of DUR that arguably meets the definition of DURC as it provides a "road-map" for the construction of a recombinant virus based on existing knowledge and techniques that could infect and transmit efficiently in human populations resulting in sterility [13]. At the time this research was published it received little attention, underscoring how difficult it is to recognize DUR, and emphasizing the need for a comprehensive and mandatory education program for DUR and relevant policies as called for in the NSABB 2007 Report.

To increase the duration of sterility from 6 months to life-long, the same researchers expressed a second gene encoding for mouse interleukin-4 (IL-4) from the genome of ectromelia virus [14]. An unintended consequence of expression of the IL-4 gene was the induction of a profound immunosuppression in the infected mouse that resulted in lethal infections of a resistant mouse strain and mice normally protected by vaccine immunity. Publication of this study in 2001 stimulated an intense debate in the popular and scientific press as the research provided a potential second "road-map" for construction of a similarly "vaccine-proof" variola virus. This second example of DUR with the ectromelia virus was discussed in the NRC 2004 Report as one of three examples of DUR.

The 2011 influenza A/H5N1 gain-of-function controversy likely provided much needed impetus for the roll-out of the 2015 DURC Policy, which focused on the 15 BSAT including the highly pathogenic avian influenza viruses such as A/H5N1. Because of the narrow focus of the 2015 DURC Policy, one of the two studies of the 2011 influenza A/H5N1 gain-of-function controversy would not have been covered by that policy even though both studies examined gain-of-function mutations in the influenza A/H5N1 hemagglutinin (H) that contribute to transmissibility of influenza virus between ferrets. The study from the Fouchier laboratory at the Erasmus Medical Center utilized a natural isolate, influenza A/H5N1/Indonesia/5/2005 virus that was genetically modified by site-directed mutagenesis of the H surface protein, and the acquisition of additional mutations by subsequent serial passage in ferrets [15]. This study would be explicitly covered by the 2015 DURC Policy. On the other hand, Kawaoka and colleagues at the University of Madison used receptor-binding studies and animal experiments to identify four mutational changes to the same H protein when expressed in a virus containing the remaining seven gene segments from a 2009 pandemic influenza A/H1N1 virus [16]. A whole 2009 pandemic influenza A/H1N1 virus or one containing the gene for the H of the more pathogenic influenza A/H5N1 virus is not covered by the 2015 DURC Policy, even though the Kawaoka virus construct was based on a 2009 influenza A/H1N1 virus that is highly communicable and few humans are thought to have protective immunity to H of influenza A/H5N1 virus. This arbitrary exclusion of one of two studies from enhanced review when both employ a similar gain-of-function approach and seek to answer the same experimental question supports the idea that all life sciences research should be evaluated at the institutional level for DURC as proposed by the NSABB 2007 Report.

There is also inconsistency in the classification of other respiratory pathogens with regard to the 2015 DURC Policy. Influenza A/H5N1 virus is included in the 2015 DURC Policy, whereas coronaviruses (CoV) severe acute respiratory syndrome (SARS) and Middle Eastern rsespiratory syndrome (MERS) are not. SARS initiated a ~9-month global pandemic in November 2002 that resulted in 8096 infections, 774 deaths and a case-fatality rate of 9.6% [17]. MERS-CoV, emerged in 2012 and as of June, 2015, has caused 1342 infections, 513 deaths, and a case-fatality rate of 38% [18]. Both SARS-CoV and MERS-CoV have infected more individuals in a shorter period of time (cumulative ~3 years) than influenza A/H5N1 virus, which is credited with 842 clinical infections, 447 deaths, and a case-fatality rate of 53% between 2003 and 2015 [19]. Both SARS-CoV and MERS-CoV are readily transmissible in human populations, while influenza A/H5N1 virus is not. These examples underscore a problem with the current USG approach for regulation of DURC. The reliance of the 2015 DURC Policy on a list of 15 BSAT suggests other potential pathogens are of less importance, and has the inherent danger of creating a false sense of security [20]. The 15 BSAT list may be expanding as in October of 2014 the USG announced a pause in DURC gain-of-function research projects (2014 Funding Pause) that would enhance the pathogenicity and/or transmissibility of influenza virus, MERS-CoV, or SARS-CoV in mammalian species by the respiratory route of infection. This pause would remain in effect

> ...until a robust and broad deliberative process is completed that results in the adoption of a new USG gain-of-function research Policy. Ref. [21]

This process would include consultation with the life sciences community, numerous stakeholders, and deliberative bodies including the NSABB. It was to be completed before October 17, 2015, the 1-year anniversary of the initiation of the 2014 Funding Pause. The 2014 Funding Pause also allowed for a future expansion of the scope of regulated research possibly to other pathogens suggesting further expansion of the "list" [21].

Thus when viewed in total, the USG policy and guidance for overseeing DURC is arguably problematic. It calls for review of life sciences research involving DURC for only a small number of agents (15 BSAT), but its reliance on a prescriptive list of pathogens can contribute to a failure to review projects with similar DURC potential, but utilizing dissimilar approaches (e.g., Fouchier versus Kawaoka studies). Equally important, the 2015 DURC Policy fails to address the need to review the vast majority of life sciences research for DURC as proposed by the NSABB 2007 Report (see also Chapter 5). DURC by pathogens not on the 15 BSAT has the potential to create public health concerns equal to that of the 15 BSAT.

A case study: IBC review of the Fouchier and Kawaoka studies that triggered the 2011 influenza A/H5N1 gain-of-function controversy

The local IBC has become the keystone of the regulatory structure known as the NIH *Guidelines* (see also Chapters 5 and 11). As noted previously, in the early years of the NIH *Guidelines*, the RAC reviewed all registration documents that involved recombinant DNA, but in 1978 the review and approval of most registration documents was conditionally delegated to the IBC. The IBC alone reviews and approves the majority of registration documents, including those involving DURC. The NIH *Guidelines* recognized that certain studies with influenza virus, such as those involving influenza A/H5N1 virus, involved increased risk to personnel and the public, and therefore required increased engineering controls, enhanced PPE, and additional practices, but not additional review. The Fouchier and Kawaoka ferret transmission studies with influenza viruses were reviewed under section III-D-7 of the NIH *Guidelines* by the IBC. For a detailed discussion of the 2011 influenza A/H5N1 controversy see Chapter 6.

Both the Fouchier and the Kawaoka groups carefully considered the potential safety and security risks associated with their studies [15,16]. The studies were reviewed for biosafety and biosecurity at many different levels, including peer-review by study section, and program review at the National Institute of Allergy and Infectious Diseases (NIAID). The IBCs evaluated the proposed research prior to commencement of the work, and assigned a biosafety level commensurate with the guidelines and regulations, and reflective of risk–benefit analyses. The projects were carried out in biocontainment laboratories that were designed specifically for influenza virus research, were staffed with well-trained personnel, and were operated at BSL-3 Enhanced, ABSL-3 Enhanced, or BSL-3Ag. For example, the Kawaoka ferret transmission experiments were carried out at BSL-3Ag, which differs from ABSL-4 only in the lack of an automatic chemical decontamination exit shower, and the use of respiratory protection based on external HEPA filtered air-supply rather than the Powered Air Purifying Respiratory (PAPR) system used in BSL-3 Enhanced, ABSL-3 Enhanced, or BSL-3Ag laboratories. In addition to the engineering controls, personnel protective equipment and practices, a robust occupational health program was in place at both institutions. Most importantly, all appropriate institutional and government approvals and inspections were obtained prior to commencement of the work. Based on the standards of the day, gain-of-function influenza A/H5N1 virus research required no additional level of review from the RAC or NIH/OBA prior to commencement of work (see also Chapter 6).

From the duration and intensity of the discussions concerning these two studies in the popular and scientific press, it is clear that some felt the IBC review to be inadequate. For this reason, it is worthwhile to examine more closely the scientific question that the research was designed to answer, and the basis of the controversy.

The first human fatality from influenza A/H5N1 virus occurred in Hong Kong in 1997. Since 2003 there have been 840 laboratory-confirmed clinical infections of influenza A/H5N1 virus and 447 fatalities, but there has been no documented, sustained human-to-human transmission. In 2006, a Blue Ribbon Panel on influenza research recommended to the NIAID that:

...evolutionary pressures that lead to emergence and spread of new viral subtypes – especially the factors that favor transmission from animals to humans – are urgent research priorities. Ref. [22]

NIAID funded Drs Fouchier and Kawaoka to undertake this line of research. Both the Fouchier and Kawaoka groups selected a gain-of-function experimental approach to search for genetic changes in the H5 that would permit efficient transmission of an influenza A/H5N1 virus between ferrets, a model of human influenza virus transmission. Of the experimental approaches available for answering this research question, gain-of-function is the most direct and time-proven. Understanding the factors that governed the evolution of influenza A/H5N1 virus transmission between humans has important implications for public health. If the research failed to identify mutations that enhanced transmission in the ferret model, it could suggest the existence of a genetic barrier to creating a highly transmissible influenza A/H5N1 virus. Influenza A/H5N1 virus infections for humans would be destined as "dead-end" interactions, as are infections with hemorrhagic disease viruses such as Lassa fever and Junin viruses. In this regard, a number of influenza virus researchers believe only influenza viruses with H subtypes of H1, H2, or H3 are capable of pandemic potential. A failure to demonstrate ferret transmissibility with mutated influenza A/H5N1 virus could support a reduction of investments in pandemic preparedness and research for influenza A/H5N1 virus. For example, worldwide at least five vaccines have been licensed against influenza A/H5N1 and more are in development. Nearly 600 million dollars and 1 billion dollars have been spent on influenza A/H5N1 vaccines in Japan and the United States, respectively.

Alternatively, if a small number of mutations could be identified that supported influenza A/H5N1 virus transmissibility in a ferret model, it could be argued that influenza A/H5N1 virus has potential to evolve into a pandemic virus, and the investment in vaccines and research in the interpandemic period is worthwhile. The identification of these mutations could provide the opportunity to develop antivirals and vaccines that would be efficacious against the "predicted" virus, and inform the influenza surveillance network. The research question was clearly important.

The gain-of-function manuscripts from the Fouchier and Kawaoka laboratories were submitted for publication and reviewed by the NSABB in late 2011 as part of its mandate to advise the USG on biosecurity issues. The NSABB recommended changes, and revised manuscripts were subsequently published in 2012 followed by a voluntary moratorium on new gain-of-function research with influenza A/H5N1 virus. The debate on the pros and cons of the Fouchier and Kawaoka studies continued unabated with the main issues concerning the risk and benefit of the research to the public, the need for an enhanced evaluation and public discussion of a small percentage of DURC, and whether public dissemination of the identity of the genetic mutations enhancing transmissibility had biosecurity implications [23].

Risk and benefit analyses of gain-of-function studies with influenza A/H5N1 virus

The risks and benefits of research are difficult to quantify, and subjective judgment plays a larger role than it should. The biosafety level assigned to an experiment is tied to the risk group of the involved agent. The greater the perceived risk of the agent to the personnel carrying out the experiment or to the public through accidental release, the greater the assigned biosafety level. Appropriate selection of the biosafety level to minimize these potential risks is important, because higher biocontainment levels are associated with increased cost and duration of the experiment (see the section, "The increasing burden of regulatory review"). Research involving influenza A/H1N1 and A/H3N2 viruses is carried out at BSL-2 as these influenza virus strains currently circulate in the human population, while research with influenza A/H2N2 virus is carried out at BSL-3 Enhanced as this strain has not circulated in human populations since 1968. Research with highly pathogenic influenza A/H5N1 virus is carried out at BSL-3 Enhanced, where the "Enhanced" indicates additional engineering controls, personnel protective equipment, and/or practices. Since the 2011 influenza A/H5N1 gain-of-function controversy there have been several publications estimating the potential public health risks associated with the study of the genetic basis of influenza A/H5N1 virus transmissibility in mammalian species. The majority of these studies based their risk estimates on data from the same publication by Henkel and colleagues, which was the first report to describe the results of the National BSAT Theft, Loss or Release reporting system for the years between 2004 and 2010 that focused on biosafety or biocontainment lapses [24]; however, as we shall see, this study lacked the necessary background information required for use of the data in robust risk analyses.

The report documented 11 laboratory-acquired infections (LAI) of which seven and four occurred in BSL-2 and BSL-3 containment laboratories, respectively. This and other studies concluded that the majority of LAI were acquired through unrecognized aerosol exposure:

> *All 11 LAIs resulted from either unrecognized and/or unreported exposures,*
> *presumably through the aerosol release of the BSATs. These observations are*
> *entirely consistent with studies by Pike [25] who found no distinguishable accidents*
> *or exposure events in more than 80% of LAIs. Harding and Byers [26] also*
> *reported only a small number of recognized containment breaks in a study of LAIs.*

The conclusion that the LAI occurred from aerosol exposure is important as the type of respiratory protection used by the affected personnel was not described in the Henkel report. The seven individuals infected in the BSL-2 containment laboratories may not have used respiratory protection as it is not required by the *BMBL* [5], whereas the four LAI from the BSL-3 laboratories would have employed no, N-95 type, or PAPR respiratory protection. The *BMBL* mandates respiratory protection only for research activities in ABSL-3 facilities, whereas the need for respiratory protection for BSL-3 facilities would be determined by the IBC risk assessment [5]. Without knowing the respiratory protection of the individuals who were infected, it is difficult to use the data to calculate probabilities of a LAI occurring as a result of influenza A/H5N1 research in BSL-3 Enhanced containment laboratories where the PAPR is the universally prescribed method of respiratory protection. Importantly, PAPR provides almost 100% protection against an aerosol infection when operating properly [27]. Similarly, the Henkel publication provided the number of individuals working in all of the reporting BSL-2, BSL-3, and BSL-4 containment laboratories, but did not attribute an individual number to each biosafety level. And, finally, influenza A/H5N1 virus experiments are carried out at BSL-3 Enhanced, which contain a number of additional safety features over BSL-3 that would minimize the frequency of exposures and the probability of a LAI.

Using this data set from the Henkel report, the derived range of probabilities for a laboratory created influenza A/H5N1 virus to cause a LAI ranged from 2×10^{-3} [28] to 1×10^{-6} [29] per laboratory per year. The 500-fold difference between the two values was due to the authors' application of various assumptions to the calculation: some included the impact of specific practices on reducing the likelihood of a LAI; others involved differing estimations of the number of labs involved in the research; and still others assumed differences in the probabilities of LAI in BSL-3 laboratories using viruses versus bacteria. Similarly, the probability that a community LAI would lead to a pandemic, and the estimated magnitude of the pandemic were equally dependent on assumptions. For comparable reasons calculating the benefit of a research project is equally difficult, and will not be discussed here.

The lack of ancillary detail to permit stratification of the data in the Henkel report suggests the derived risk analyses may be dominated by subjective assumptions, and therefore are of reduced value in determining the probability of a LAI originating from a BSL-3 Enhanced laboratory carrying out gain-of-function research with influenza A/H5N1 virus. Further, the IBCs evaluated the Fouchier and Kawaoka studies prior to publication of the Henkel report. Thus, even these limited data may not have been available for risk–benefit analyses, and to inform decisions on whether additional biosafety or biosecurity measures were necessary and would be meaningful.

The lack of stratified exposure and LAI data could be mitigated to some degree, if reports describing high-containment biosafety laboratory exposure data described the detailed biocontainment context (e.g., detailed descriptions of engineering controls, personnel protective equipment, and practices) in which individuals worked, were inadvertently exposed to an agent, and at a low frequency became infected. In order to continue to improve best practices for biocontainment and biosecurity, it is especially important to have accidental exposure data (or the lack thereof) from the new state-of-the-art BSL-3/4 National Biodefense Analysis and Countermeasures Center and Integrated Research Facility at Frederick, MD, and from similar facilities in other countries. And, finally, along with better data, there needs to be agreement in the scientific community on the appropriateness of the assumptions that are used in calculating risk–benefit analyses for DURC. Together this will enable the IBC to evaluate better the DUR projects it reviews.

Options for an additional level of review for DURC

Current life sciences research involving pathogens and recombinant DNA is overseen by a series of biosecurity laws, Executive Orders, administrative orders, and guidance documents. The NIH *Guidelines*, 2013 DHHS framework, and 2015 DURC Policy inform the actual manner of the review process for some, if not all, life sciences research. The NIH *Guidelines* delegate review and approval of registration documents conditionally to the IBC. The NIH *Guidelines* require certain experiments that potentially involve DURC (Table 13.2) to require additional levels of review. For example, the deliberate transfer of a drug resistance trait to a certain microorganism could require IBC review, RAC review, and NIH Director approval prior to initiation of the study (NIH *Guidelines*, Section III-A). Also, the cloning of a toxin molecule with LD_{50} of less than 100 ng/kg requires IBC and NIH/Office of Biotechnology Activities (NIH/OBA) approval (NIH *Guidelines*, Section III-B). The NIH **Guidelines** also recognize that certain influenza virus studies involving genes from 1918–1919 H1N1 (1918 H1N1), human H2N2 (1957–1968), and influenza A/H5N1 viruses require special consideration (enhanced biocontainment), but not additional review. Thus, as mentioned previously, the Fouchier and Kawaoka gain-of-function studies needed only to be reviewed at the level of the IBC. This hierarchical review approach for recombinant and synthetic DNA experiments has worked well for almost 40 years.

The current 2015 DURC Policy calls for the evaluation of DUR involving the 15 BSAT at the level of the Institutional Review Entity (IRE), which most often will be the IBC (for detailed discussion of the mechanics of DURC review see also Chapter 6). The IRE uses a risk–benefit assessment analysis to determine whether proposed research is DURC. If the research is determined to be DURC, the PI, working in conjunction with the IBC, would draft a risk mitigation plan to guide the conduct and communication of the DURC. The risk mitigation plan would need to be approved by the USG funding agency. This approach raises at least three questions. First, would using the agency that funded the project to review a risk mitigation plan be a conflict of interest? Second, what individual(s) or group of individuals would carry out the review? For example, would the program officer operate in another capacity to review the risk mitigation plan or would it be someone who was trained in risk-benefit analysis? And third, how would independent funding agencies ensure consistent reviews of mitigation plans for an identical DURC?

In addition to the 2015 DURC Policy, funding proposals for influenza A/H5N1 gain-of-function studies must also be responsive to the 2013 DHHS framework. Under this framework an influenza A/H5N1 virus gain-of-function research proposal would be reviewed for Scientific Merit and dual-use issues by the funding agency. If the proposal is in the fundable range, and passes the seven-point DHHS criteria, it would be forwarded to DHHS for a further rigorous review based on scientific and public health benefits, biosafety and biosecurity risks, and risk mitigation plan. If the research project passes the DHHS review process, the funding agency is approved to send the PI a notice of award. The 2013 DHHS framework also raises numerous questions, three of which are listed here. Would the funding agencies require an IBC reviewed registration document, including a risk–benefit analysis and risk mitigation plan, to accompany the funding solicitation? This would make sense as otherwise the DHHS review committee would have to generate *de novo* the risk–benefit analysis and the risk management plan for each project it reviewed. Although perhaps advantageous for the DHHS review committee, this approach would increase the IBC workload as most IBCs review registration documents only after the award of funds as ~80–90% research proposals are not funded. If the DHHS review committee did generate its own risk–benefit analysis, would the IBC be able to modify a DHHS-approved risk mitigation plan or would the DHHS risk–benefit assessment replace that aspect of the IBC review? And, finally, would the DHHS risk mitigation plan replace the need for the risk mitigation plan mandated in 2015 DURC Policy, since the plans likely would end up in the same place, that is the funding agency? It is not clear how this will work in practice, but the PI and IBC will have additional responsibilities and reporting requirements when these policies and guidelines take effect. This will likely make the PI less competitive with researchers abroad who are not encumbered by this regulatory burden.

The NRC 2004 Report and NSABB 2007 Report describe seven similar categories of DURC that require enhanced oversight. The NSABB 2007 Report proposed the evaluation of all life sciences DUR by an IRE, possibly the IBC with expanded expertise, and emphasized the role of this local oversight in managing DURC. DURC would be identified by risk–benefit analysis using a suite of tools described by the NSABB [10]. Identified DURC would be managed by a risk mitigation plan developed by the IBC and PI, with no additional level of review described. The NRC 2004 Report proposed review of DUR involving BSAT at the level of the IBC. For most DURC an acceptable risk management plan would be developed between the PI and IBC, however, the authors of the NRC 2004 Report recognized that certain DURC would have a greater degree of inherent risk, and would require an additional level of review. Based on the success of the NIH *Guidelines* at facilitating research while protecting public safety, the NRC 2004 Report proposed a similar hierarchical review process for evaluating DURC. The vast majority of DUR involving BSAT would be handled at the level of the IBC, but some or all of the experiments in the seven categories as defined in the 2015 DURC Policy and listed in Table 13.2:

> *…would be referred to an expanded RAC and possibly for approval or denial of permission to proceed with the proposed experiment. Ref. [6]*

This approach has the added benefit that all DURC categories except #2 are currently required by the NIH *Guidelines* to be reviewed by the IBC (Table 13.2), with certain experiments in categories 3 and 7 being further reviewed by the RAC or NIH/OBA. Importantly, unlike 2015 DURC Policy, the NIH *Guidelines* require IBC review of all experiments that involve DURC in these categories not just experiments involving the 15 BSAT.

The NIH *Guidelines* would need updated criteria to determine which DURC would be reviewed solely by the IBC and which DURC would require secondary review by an expanded central or regional RAC. These criteria could be developed by the

NSABB in consultation with the scientific community. Using an expanded RAC to review selective DURC would have a number of advantages over current approaches or that proposed in 2015 DURC Policy:

1. One national entity would review registration documents ensuring consistency of review.
2. The RAC already reviews certain category 3 experiments that involve the transfer of a drug resistance trait to certain microorganisms, so there is already an "institutional experience" with the review of DURC.
3. It would minimize the potential for conflict of interest.
4. It would improve the effectiveness of the IBC and improve the quality of the review of DURC. Under this scenario the DHHS review committee would review scientific and public health benefits as it pertained to its research portfolio, but biosafety and biosecurity risks and the generation of a risk mitigation plan would be left to the IBC and the expanded RAC.

Prior to 2011, the IBC had the major role in the evaluation of influenza A/H5N1, SARS-Cov, and MERS-CoV gain-of-function research proposals for biosafety and biosecurity based solely on the NIH *Guidelines*. The reaction to the 2011 influenza A/H5N1 controversy has been an inadvertent reduction of the IBC role in the management of this class of experiment, the addition of more oversight to the review process, and the potential for expansion of the "list" with the addition of more pathogens and more classes of DURC (2014 Funding Pause). The current reaction to gain-of-function DURC with a quantitatively unknowable level of risk is to impose additional layers of oversight and/or layers of biosafety and biosecurity. This may be justifiable for a very small number of DURC, but in our view not the vast majority. Although harder to document, an emphasis on education and proficiency training of those involved in life-sciences research, with special attention to individuals involved in DURC, may have a greater impact at lowering risk without impacting science (see the section, "The increasing burden of regulatory review"). In summary, we argue that the local IBC, if provided with increased resources, is capable to competently review all registration documents for DURC and to manage the vast majority, itself, while forwarding only a select few to an expanded RAC for secondary review.

The need to expand the IBC's role in life sciences research

Since the IBC assumed the role of primary reviewer of recombinant DNA research in 1978, it has had a dramatic increase in its responsibility to oversee biological research and to engage in biosecurity as well as biosafety [30] (see also Chapters 5 and 11). The increased responsibility is based partly on recent USG policy expanding its review role for DURC with the 15 BSAT. Our view is that IBC review of DURC should be expanded even further than mandated by the 2015 DURC Policy to include all of life sciences research. Although this would further increase the workload of the IBC, it would result in a more comprehensive picture of DURC in life sciences research, especially research carried out in laboratories that have low to no biosafety containment and biosecurity. An example of this type of research leading to DURC is the immune-contraceptive ectromelia-ZP-GP-3 recombinant and the vaccine proof ectromelia-IL-4 recombinant described above. IBC expanded review would also eliminate the gaps in current 2015 DURC Policy by regulating research with all pathogens with pandemic potential, not just pathogens currently on a "list." The DUR dilemma is now at least 25 years old, and although it has not been adequately addressed, in many ways it is an old debate. The IBC is also well-positioned to carry out additional activities as warranted by the changing landscape of science.

For instance, the IBC is in an ideal position to foster outreach with the local community as there is a number of important topics with which to engage the public (see also Chapter 7). The 2011 influenza A/H5N1 gain-of-function controversy has raised a concern that certain DURC experiments are too risky to carry out due to low probability but highly consequential event, such as an LAI leading to an influenza virus pandemic. The IBC is well suited to explain to local communities the review process, including the risk–benefit assessment that is a component in the evaluation of DURC. A series of USA Today articles have suggested that not all of the nation's biocontainment laboratories are being operated in a safe and secure manner [31]. This comes at a time when trust between the scientific establishment and the public is already strained due to the inadvertent shipment of live *Bacillus anthracis* [32] and live influenza A/H5N1 virus [33] from CDC laboratories; the shipment of *Bacillus anthracis* from the Army's Dugaway Proving grounds in Utah over a 12-year period to at least 194 laboratories in all 50 states, Washington, DC, Guam, the US Virgin Islands, Puerto Rico, and nine other countries [34]; and the release of *Burkholderia pseudomallei* from a high-containment BSL-3 laboratory at the Tulane National Primate Research Center [35]. It is important to note that the majority of these mishaps occurred at USG laboratories, and not at academic institutions that receive the vast majority of USG life sciences funding. The IBC is well suited to reassure local communities of the safety and security of the life sciences enterprise, as well as its importance for homeland security and for future gains in human medicine and agriculture.

Recent innovations have opened up exciting areas of research (see also Chapter 7). In the last few years, a new technology called CRISPR (Clustered Regularly Interspaced Short Palindromic Repeat) has been evolving at a tremendous pace in research laboratories around the world. The technology permits modification of DNA of humans, other animals, and plants such that transcriptional regulation can be altered or genes added or subtracted [36,37]. Importantly, compared to previous techniques for modifying DNA, this new approach is much faster and easier, and promises to change the manner in which certain diseases are treated. There is already at least one report of Chinese researchers modifying human embryos [38]. In addition, synthetic biology offers the possibility to modify microorganisms by the addition of functional "genetic circuits" and metabolic pathways for such practical purposes as the production of pharmaceuticals and biofuels, to break down pollutants, and, in the longer term, to create new life-forms [39]. And finally, in agriculture the continual increase in the world's population under changing environmental conditions and decreasing water supplies will drive the plant biotechnological revolution to generate new solutions for food security (see Chapter 9 for crop-specific IBC issues) [40]. All of these new areas of research will require a certain level of oversight that the IBC is best positioned to provide.

Although the responsibilities of the IBC should continue to expand at most institutions, it is ill-prepared for the future. As compared to the Institutional Animal Care and Use Committee and the Institutional Review Board, the IBC receives a fraction of the resources, has equal or greater regulatory burden, and is responsible for a broader swath of science [41]. The USG needs to determine a mechanism to directly fund the IBC without leaving this task to the institution (see section, "Regulatory compliance for research institutions: an unfunded mandate").

Education and training of the life sciences scientists: PIs and IBC members

We propose that all of life sciences research be reviewed for dual-use, and the small percentage of studies with DURC that meets a particular criterion would require an additional level of review by an expanded RAC. For this system to work effectively, the PI that prepares the IBC registration document will self-identify DUR that may be considered DURC. For its part, the IBC will need to separate DURC into projects that can be evaluated completely at the level of the IBC, and those needing secondary review by the expanded RAC. This process must be consistent among IBCs. Ideally, projects with similar DURC would be managed by similar risk management plans, even at different IBCs. Also, it would be advantageous for the local IBC to be staffed with the required subject matter experts such that each registration document receives a thorough evaluation. This could be difficult for some smaller institutions, and it would be important to have an expanded availability of commercial, virtual IBCs and/or a national register of trained *ad hoc* IBC members. Equally important, the PIs and members of the IBC must receive education and training on DUR, especially risk–benefit analyses.

Much has been written about the need to improve the education and training of life scientists. Both the NRC 2004 Report and the NSABB 2007 Report highlight the importance of education of the life sciences community on DUR issues. The NSABB further published a report entitled: "Strategic Plan for Outreach and Education on Dual-Use Research Issues" [8]. This report described a comprehensive educational outreach strategy targeting multiple target audiences (e.g., Congress, general public, scientists, laboratory staff, students/trainees, research administrators, and institutional leadership) through multiple venues (e.g., professional associations, scientific societies, scientific journals, opinion leaders, and the popular press). The Office of Biotechnology Activities in the Office Scientific Policy at NIH sponsors a comprehensive website that provides education materials on DUR [42]. More recently, the 2015 DURC Policy directed that institutions must provide education and training for individuals conducting life sciences research with the 15 BSAT. In addition to these educational materials, members of the NSABB engage target audiences on the issue of DUR, and the Office Biotechnology Activities presents "NIH Guidelines 101" at the American Biological Safety Association's annual meeting and at other venues.

In spite of all of these educational efforts the impact on the life sciences researchers appears to have been modest. There is a low level of interest and knowledge of the dual-use dilemma shown by senior scientists and institutional officials [43]. The majority of IBC members, PIs, and laboratory staff have not received training on DUR [44]. And, most importantly, there is little federal funding available on a direct cost basis for education of dual-use issues [43]. It is not clear how this situation will change until USG policy requires education and training on DUR issues of all individuals involved in life sciences research at the institutional level.

The increasing burden of regulatory review

As mentioned above, there are very little data addressing the burden of IBC compliance and/or review on research, although there is a small literature on the more general question of regulatory burden on faculty workload. In a 2002 paper, Skorton et al. discuss the tensions that have arisen between the two missions of academic research, namely, promoting and supporting a vigorous research agenda while regulating the very programs that these activities comprise [45]. Focusing largely on human and animal subjects' protection, the article presents a call to action, to consolidate regulatory activities, streamline duplicative and unnecessary reporting, and reduce costs. Other similar discussions in the literature are available [41,44,46–50]. Few would argue that any progress has been made in reining in regulatory burdens during the 13 years that followed this plea. In fact the problem appears to be getting worse. A 2014 National Academy of Sciences report found PIs spent almost half of the time assigned for federally funded projects on meeting the requirements imposed by those projects. As many as 23 separate administrative activities – both pre- and post-award – were identified [51].

A nationwide survey of life sciences researchers, conducted in 2009 by the Center for Biodefense, Law and Public Policy, examined the degree of anxiety over inadvertent violations of BSAT regulations [52]. The survey tested the hypothesis that BSAT should *not* be regulated. Despite uncertainty regarding the current structure and dissemination of the guidelines and regulations, the survey showed that scientists were overwhelmingly in favor of continued regulation. Of particular interest was the observation that many researchers were more concerned about committing a violation than they were about personal injury. These data present a disturbing picture of the kinds of pressure (both personal and regulatory) that many users of the IBC may be experiencing, although the date of the survey (2009), only 5 years into the federal BSAT program, leaves room for the possibility that IBC regulations have since been clarified and attitudes have adjusted.

Dias et al. examined the effects of the USA PATRIOT Act and the 2002 Bioterrorism Preparedness Act on BSAT research in the US [53]. In 2010, the most salient programmatic observation was a measurable attrition from the field of the study topic (*Bacillus anthracis*) compared with the control topic (*Klebsiella pneumoniae*). The results indicated that before 2002, the average number of papers dealing with *Bacillus anthracis* was 17; that number declined to three by 2006. In contrast, a similar accounting of papers focused on *Klebsiella pneumoniae* found a smaller decrease (from 26 to 17 research papers per millions of dollars of US funding). Thus, BSAT research activity dropped by 65%, while general pathogen research dropped by only 17%. While the authors recognized a number of problems with the analysis (so that the numbers may be "soft"), the data clearly indicated that there was a 2–5-fold decrease in certain BSAT research, as estimated by the number of BSAT research papers published in the United States.

There have been a number of publications discussing the efficacy of the current BSAT program and the related ability of the IBC to contribute to biosecurity and biosafety, through such avenues as dual-use review, inventory, personnel reliability, SOP evaluation, etc. [20,48,49,54,55]. The potential contribution of the IBC to these kinds of biosecurity issues is discussed in other chapters of this volume (see Chapter 11). In a novel approach to relieving the burden of IBC-related responsibilities, Bosselman et al. [46] addressed the problems of staff turnover and training requirements, anxiety, duplication of records and risk assessments, etc., by analyzing specific issues that were causing the most burden and delay. They found major obstacles in IBC protocol management (coordinating correspondence between investigators and reviewers), in managing training information, and in recordkeeping for laboratory inspections. Their solution was to create a collaboration platform that organizes and displays data, documents, and tasks in a coordinated structure, so that the processes required for progress could be tracked simultaneously.

Morse [48] describes the evolution and implementation of pathogen-related regulations and outlines the discussion within the research community of their effect on productivity, international collaboration, and cost. Morse points out that "many of the proposed negative impacts of the[se] regulations are based on anecdotal evidence." For example, a number of microbial collections were destroyed by investigators when the BSAT regulations came into effect, and the BSAT Program tried to prevent further losses by moving culture collections to registered laboratories [20]. Many researchers complained that they had to use research funds to upgrade their laboratories to appropriate security standards [20]. Gaudioso and Salerno bemoan the loss of personal and institutional talent applied to BSAT research, and relate the flight from the field directly to the burdensome nature of registration and compliance; more than one high-profile researcher announced that they would change their research focus, rather than submit to the regulations [54]. BSAT requirements have grown only more restrictive and demanding in the ensuing years. Sutton [50] argues that researchers are the sole targets of current biosafety and biosecurity regulations; indeed, Sutton develops the argument that the alleged perpetrator of the anthrax attacks of 2001 would not have been detected under (then) current BSAT Program, and adds:

It may be time to rethink the regulatory framework for the nation's biodefense research, focusing more on fully implementing the background investigation requirements and better perimeter security, and placing less emphasis on filling out forms, filing reports and counting units of self-replicating organisms (2009).

With the recent implementation of the Personnel Suitability Assessment program under the Federal BSAT Program (2012), a program intended to determine whether a person who intends to work with (tier 1) BSAT actually warrants access, the attendant bureaucratic responsibilities have only increased. A recent report reviewed trends in IBC practices by evaluating responses to surveys of IBCs registered with the CDC from 2002, 2007, and 2010 [44]. The data were mixed: incident training and incident reporting remained problematic, as were interactions with other regulatory committees (Institutional Animal Care and Use Committees (IACUC) and Institutional Review Boards (IRB)), while improvements in management, staffing, and compliance improved over the 8 years covered by the surveys. But Jenkins points out, the Hackney study, while useful in analyzing general trends in the development of the IBC, lacked "quantitative data on the increasing registration and policy burdens of IBCs over time, as well as on the number of protocol reviews, administrative, resources, and financial support" [41].

Defense against biological threats, both natural and intentional, maintains a position of high priority in basic research, public health, and the clinical enterprise. The security implications of the spread of infectious disease have not gone unnoticed at the national and global levels. Furthermore, the impact of the anthrax attacks of 2001 has been felt throughout the global community. These and other events have significantly transformed the landscape of the basic science enterprise – particularly in the field of infectious disease. This discussion of the burdens associated with BSAT and infectious disease research leaves a number of questions unanswered. How can we maintain safe and secure research facilities while conducting research efficiently? How can we factor in the impacts of research policies on research quality and productivity? How can we find a "healthy balance between facilitating research and protecting against audit and legal concerns [51]? Going forward, the IBC will play an increasingly central role in the increasingly complex regulatory environment associated with life sciences research. Reliable quantification of the burden of IBC requirements on the research enterprise remains unavailable. In order to exploit the expertise and other contributions offered by a well-informed IBC, the burdens to be placed on these committees warrant systematic review.

Regulatory compliance for research institutions: an unfunded mandate

As documented in previous chapters and here, the scientific community has collaborated actively with the federal government in recognizing and minimizing the risks of DURC. It is equally obvious, however, that these measures, individually and collectively, require research institutions to commit substantial resources, for which those institutions assume full financial responsibility. The issue raised by these developments is not whether institutions should or should not assume such responsibilities if they want to conduct DURC. Instead, the issue is whether institutional resources are adequate to ensure the safe and secure conduct of DURC, especially at a time when many research institutions face increasing financial stress related to the conduct of research. While those stresses are a function of several trends that apply generally to federal funding, they have clear implications for DURC: (i) decreasing levels of funding for biomedical research over the past decade, in actual dollars and as a percentage of the federal budget, and (ii) an inexorable increase in federal regulations since 1991, the year when the current cap on administrative costs on federal awards was established, that exacts a cumulative financial burden for regulatory compliance. The following sections consider each of these issues in more detail.

Federal funding trends

Federal funding for biotechnology research has long been critical to scientific advances that, in turn, have contributed to improved public health and to a growing economy [6]. That funding also has occasioned growth in the biosciences: NIH funding in 2013 supported over 300,000 researchers at over 3000 universities and research institutions [56], reflecting the critical role of NIH funding for the biomedical science community. While a longstanding and critical support for the bioscience enterprise, federal research funding, as a proportion of the federal budget, has declined steadily over the past half-century; it now accounts for less than 2% of that budget (see Figure 13.1). Similarly, in real dollars, funding for biomedical research (as reflected in the NIH budget) also has declined. Over the past decade, NIH Research & Development funding (in constant 2015 dollars) has declined from $32,068 million in

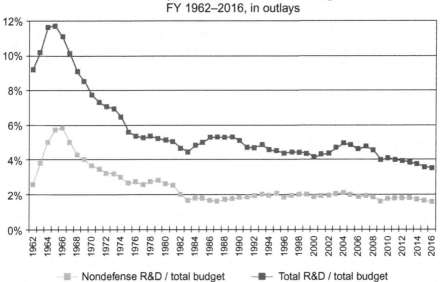

R&D as percent of the federal budget:
FY 1962–2016, in outlays

Nondefense R&D / total budget   Total R&D / total budget

Figure 13.1 Research funding as a percent of the federal budget, 1962–2016.
Source: Budget of the US Government FY2016 historical tables. FY2016 is the request.
© 2015 AAAS.

FY2007, to $29,182 million (requested) in FY2016 [57], a decline of ~9%. Moreover, the trend is projected to continue in the aftermath of the 2013 sequester and ongoing political squabbling over funding for domestic programs.

Paradoxically, this decline in funding comes at a time when the number of applications for funding has increased. An examination of applications from FY2005 to FY2014 (the most recent data for which success rates are available), shows that all-agency applications increased by 18.6%, from 43,069 to 51,073 [58]. As a consequence of the increasing applications for a decreasing pool of funds, success rates for NIH awards also have declined over that period, from 22.3% in 2005 to 18.1% in 2014. These and other developments have boosted costs borne by research institutions, secondary to several factors [56]:

1. escalating costs of faculty recruitment,
2. the costs of establishing viable research laboratories,
3. startup costs for new research programs,
4. the cost of bridge funding for established researchers who confront a hiatus in external funding, and
5. the use of internal funds to support research agendas that, eventually, could attract external funding.

This increase in institutional research costs is reflected in FY2010 data collected by the National Science Foundation through their annual Higher Education Research and Development (HERD) Survey, which shows that institutional funds directly supporting research totaled $9.1 billion [59]. When combined with institutional support for associated "indirect" (i.e., Facilities and Administrative) expenses, institutional funding in 2012 accounted for 21.6% of all Research & Development expenditures, compared to 18.0% of R&D expenditures in 2006. Hence, institutional R&D funding in 2012 was second only to that provided by federal sources (59.5%). Moreover, escalating institutional compliance requirements, such as those associated with DURC, suggest that the financial burden on institutions to provide internal research funding will continue to increase, a trend that can threaten the financial ability of all but the strongest research institutions to support the research enterprise.

Costs of compliance

As noted above, research institutions face rising costs associated with long-term trends that place increased pressure on both direct and indirect research-related expenses. Both direct and indirect costs of research represent real institutional expenditures. Direct costs, of course, are those elements that are readily identified (and represent the bulk of an award) in a research project, such as supplies, research equipment, investigator salaries, etc. Indirect costs are not so clearly linked to a given project, but include expenses for facilities, infrastructure, and operational activities (e.g., electricity, heat, ventilation, maintenance, disposal of hazardous waste), as well as costs associated with the administration of awards (e.g., compliance with regulations, cost controls and financial reporting, information security). In the case of research involving BSAT, there are additional indirect costs involved with security systems and personnel, maintenance of specialized engineering controls such as HVAC systems, administrative panels to evaluate the suitability of research-related personnel ("personnel assurance programs"), and the requirement for a Responsible Official to oversee program administration (see also Chapters 3 and 6). The impetus for the federal government to set Facility and Administration (F&A) rates, of course, was to support the indirect costs associated with the conduct of federally funded research. Hence, F&A rates are negotiated by institutions that receive federal funding, typically via audits that are undertaken (at an institution's expense) every 3 years. While an institution's F&A rate is subject to change (either up or down) over time, the maximum amount of administrative expenses that an institution can claim is capped at 26% of direct costs of a federal award, a rate set in 1991.

Regulatory changes since 1991

1991 was a significant year for federal oversight of research, as it not only was the year in which formulas were set for F&A rates, but also was the year when the Common Rule was adopted to provide for human subject protection across federal agencies. While the cap on administrative cost rates is unchanged since 1991, the same cannot be said of the regulatory landscape that contributes to those costs. Recently, the Council on Government Relations (COGR) compiled a list of regulations that have been implemented or amended between 1991 and 2013, as well as interpretations/ implementations that have impacted business practices [60]. That document lists 52 regulations/amendments, 20 interpretations/implementations, and nine further regulatory changes that were proposed at the time that the document was published. A partial list of regulatory requirements follows with which most investigators will be familiar, as many impact the management of federally funded awards:

- 1994: NIH Guidelines for Research Involving Recombinant DNA Molecules
- 1994: Deemed Exports (EAR & ITAR)
- 1995: Conflicts of Interest (Amendments August 2012)
- 1996: Health Insurance Portability & Accountability Act of 1996 (HIPAA) Privacy Rule (Amendments January 2013)
- 2000: Misconduct in Science (Federal-wide Policy)
- 2002: BSAT (under CDC and USDA/APHIS) Public Health Security & Bioterrorism Preparedness & Response Act of 2002; companion to the USA PATRIOT Act (2001); revised October 2012
- 2002: FISMA: Federal Information Security Management Act (Title III, E Government Act of 2002) OMB Circular A-130, Management of Federal Information Resources, Appendix III, Security of Federal Automated Information Systems
- 2008: Homeland Security Chemical Facilities Anti-Terrorism Standards (CFATS)
- 2009: National Institutes of Health Guidelines for Human Stem Cell Research
- 2006: National Science Foundation Responsible Conduct of Research Training (America COMPETES Act; implemented 2010)
- 2012: US Government Policy for the Oversight of Life Sciences Dual Use Research of Concern
- 2013: NIH, Mitigating Risks of Life Science Dual Use Research of Concern.

Faced with the ongoing need to comply with these and other regulatory and/or financial reporting requirements, research institutions have developed multiple structures to manage these recurrent functions. This is reflected to some degree in compliance committees, many involving faculty representation, that include IRB, IACUC, IBC, IRE, Conflict of Interest (COI) committees, etc. While those committees largely depend on the willingness of research faculty to volunteer time in support of committee functions, the infrastructure behind these (and other) committees cannot be supported on a voluntary basis. Indeed, that infrastructure requires expert personnel, information systems, office space, and other institutional resources.

Because of the multiplicity of these operations and the varying organizational structures that support them, it is very difficult to obtain an exact accounting of the institutional costs exacted by these compliance-related operations. That said, several recent studies provide estimates of costs associated with the array of research-related regulatory requirements associated with federal funding. The previously referenced HERD study estimated that indirect costs incurred by research institutions in FY2010 exceeded federal F&A support by $4.6 billion dollars. Another survey, undertaken by the Association of American Medical Colleges (AAMC), examined indirect costs of medical schools with research programs that received federal funding that ranged from moderate ($26 million) to very high ($751 million) levels in FY2013 [56]. That study showed that institutional expenditures for unreimbursed indirect costs of external awards averaged $0.15 per dollar of direct costs. As noted previously, other internal expenditures also were incurred (e.g., startup packages, additional salary support for research effort, bridge funding), so that the combined institutional costs amounted to $0.53 per dollar of direct costs.

Of course, the costs of compliance are not only measured in dollars nor are they visited only on research institutions. Those costs are visited on investigators, as well, often in the form of reduced research productivity. Data regarding reduced productivity have been collected by the Federal Demonstration Partnership (FDP), a collaboration among 10 federal agencies and 119 federally funded institutional partners that is sponsored by the National Academies [51]. The FDP survey was completed by 13,453 Principal Investigators with active federal grants in FY2010 (a 26% response rate). Respondents identified many administrative functions that consumed their effort (see Table 13.3), greatly reducing (to 57.7%) the estimated proportion of funded effort actually spent on active research.

The way forward

Biomedical research institutions are under increasing financial pressure in the face of declining levels of NIH funding and accumulating administrative costs associated with ever-increasing regulatory and financial reporting requirements. While the latter trend affects all research institutions, it might be argued that it particularly affects those engaged in DURC, as the safety and security requirements for that form of research are particularly steep. In fact, the requirements are sufficiently demanding that some research institutions have moved away from DURC research because of the costs of continuing such investigations relative to the costs of other forms of

Table 13.3 Investigator administrative requirements associated with federally funded research

General research administration
Personnel: Personnel administrative issues (hiring, managing, visas, evaluation)   *Finances*: Managing grant/contract expenditures   *Effort Reporting*: Federal time and effort reporting, including training   *Data Sharing*: Meeting federal requirements for resource and data sharing   *Cross-Agency*: Dealing with differences in administrative requirements across federal agencies
Compliance
COI: Meeting federal conflict of interest requirements   *HIPAA*: Meeting Health Insurance Portability and Accountability Act requirements   *IACUC*: Meeting federal animal care and use requirements   *RCR*: Meeting Responsible Conduct of Research requirements for trainees on federally funded projects
Safety/security
Biosafety: Managing biohazards and bloodborne pathogens   *Chemical Safety*: Chemical inventory use and management   *Controlled Substances/Narcotics*   *Export Controls*   *General Laboratory Safety/Security*   *Radiation Safety*   *Recombinant DNA*   *Select Agents*
Other
Clinical trials: Responsibilities associated specifically with conducting clinical trials   *Intellectual Property*: Patient/copyright applications, licensing agreements, invention disclosures, Material Transfer Agreements, etc.   *Subcontracts*: Responsibilities associated with managing subcontracts to other entities [51]

biomedical research [42]. Hence, relative to biomedical research in general and to DURC specifically, we may be approaching a "tipping point" [60]:

> *A decline in the quality of research infrastructure and compliance oversight,*
> *a gradual degradation of laboratories and facilities, and ultimately, lost*
> *competitiveness...to conduct research.*

These strains on research institutions are of particular note in light of current bio-terrorism threats, which underscore the importance of scientific breakthroughs that can mitigate the effects of such threats. Indeed, Congress has recognized both the threat and the importance of scientific breakthroughs as part of its national strategy for responding to bioterrorism [43]. Unfortunately, the effectiveness of the scientific arm of the "war on terror" can be undercut by the converging stresses discussed above, which shift increasing costs of research onto research institutions. Although the cost shift may be inadvertent, its magnitude is substantial and its long-term effects prejudicial to the health of the US scientific enterprise at a time when the potential for truly significant breakthroughs is at historically high levels.

Several recent reports have targeted the unfunded mandate associated with the regulatory burden as a major contributor to this tipping point. For example, one report addressed the broad challenges facing research universities [44]. Of the 10 recommended actions, several were aimed at mitigating the regulatory burden described above: (i) that federal awards fully fund the direct and indirect costs of research (recommendation 6), and (ii) that the federal government reduce costly regulations that do not substantially improve the research environment (recommendation 7). Similar themes were echoed in another report that advanced more specific solutions [45]:

1. to eliminate regulations that do not add value or enhance accountability;
2. to build flexibility into regulations so that they allow appropriate adjustment of rules (e.g., current Chemical Facilities and Anti-Terrorism Standards require that universities apply policies and procedures that are identical to those imposed on chemical manufacturers); and
3. to apply the Unfunded Mandates Reform Act more broadly to universities, rather than restrict it to state/local governments and tribal entities, so as to account for unfunded costs of new regulatory requirements.

Aside from the proposals described above that target the costs of the broad regulatory burden faced by research universities (which we endorse), a number of additional steps could be taken that could benefit DURC programs. Indeed, a range of suggestions emerged from a 2013 conference aimed at tempering some of the negative effects of the 2012 BSAT Regulations [42]:

1. The USG and research institutions should jointly develop a new, systematic approach for promoting, supporting, and overseeing BSAT research.
2. The USG should provide a funding mechanism to support maintenance of existing facilities and infrastructure upgrades (both physical and personnel security efforts) to meet the new security requirements for tier 1 BSAT.
3. The USG BSAT Program should prepare letters of interpretation or frequently asked questions for the most prevalent and/or concerning security findings encountered during inspections, particularly when inconsistent findings have arisen.
4. Increased training, communication, and flexibility amongst regulators and between institutions are needed to introduce consistency into the inspection process and uniformity in the implementation of the regulations.

Biotechnology advances have spurred rapid growth in our understanding of biological processes and in our ability to influence those processes. In one of the great ironies of our times, these advances carry not only enormous opportunities for discoveries that can improve human health, but also opportunities for such discoveries to be used as weapons. Balancing the benefits of such research against its risks is a critical undertaking for the scientists who make the discoveries, the institutions that support their work, and the society whose members endorse the scientific endeavor. In recent years, the costs of maintaining that balance have grown and are visited disproportion-ately on the institutions in which the scientific work occurs. For these advances to continue, a rebalancing of the historically successful university–federal partnership is needed. We believe that the changes proposed in this chapter could rebalance that partnership and allow research institutions to remain viable enterprises, accommo-dating both scientific discoveries that can benefit human health and the precautions necessary to ensure the safety and security of the broader society.

References

[1] Petsko GA. Hypothesis-driven genomics pays off. Genome Biol 2012;13(10):176.

[2] Berg P, et al. Summary statement of the Asilomar conference on recombinant DNA mol-ecules. Proc Natl Acad Sci USA 1975;72(6):1981–4.

[3] Berg P, et al. Potential biohazards of recombinant DNA molecules. Science 1974;185(4148):303.

[4] NIH guidelines for research involving recombinant or synthetic nucleic acid molecules (NIH Guidelines). National Institutes of Health, 2013. Available from: <http://osp.od.nih.gov/sites/default/files/NIH_Guidelines_0.pdf> [accessed 10.07.15].

[5] Wilson DE, Chosewood LC, editors. Biosafety in Microbiological and Biomedical Laboratories (BMBL) (5th ed.). U.S. Department of Health and Human Services, 2009. HHS Publication No. (CDC) 21-1112. Available from: <http://www.cdc.gov/biosafety/publications/bmbl5/> [accessed 10.07.15].

[6] Committee on Research Standards and Practices to Prevent the Destructive Application of Biotechnology, National Academy of Sciences, National Research Council. Biotechnology Research in an Age of Terrorism. Washington, DC: National Academies Press, 2004.

[7] NSABB. Strategies to educate amateur biologists and scientists in non-life science dis-ciplines about dual use research in the life sciences. Washington, DC: NSABB, 2011. Available from: <http://osp.od.nih.gov/sites/default/files/resources/FinalNSABBReport-AmateurBiologist-NonlifeScientists_June-2011_0.pdf> [accessed 11.07.15].

[8] NSABB. Strategic plan for outreach and education on dual use research issues. Washington, DC: NSABB, 2008. Available from: <http://osp.od.nih.gov/sites/default/files/resources/Strategic%20Plan%20for%20Outreach%20and%20Education%20on%20Dual%20Use%20Research%20Issues.pdf> [accessed 11.07.15].

[9] NSABB. International engagement on dual use research. Available from: <http://osp.od.nih.gov/office-biotechnology-activities/biosecurity/dual-use-research-of-concern/international-engagement> [accessed 14.07.15].

[10] NSABB. Proposed framework for the oversight of dual use life sciences research: strategies for minimizing the potential misuse of research information. Washington, DC: NSABB, 2007. Available from: <http://osp.od.nih.gov/sites/default/files/biosecurity_PDF_Framework%20for%20transmittal%200807_Sept07.pdf> [accessed 14.07.15].

[11] United States Government Policy for Institutional Oversight of Life Sciences Dual Use Research of Concern. US Department of Health and Human Services, 2014. Available from: <http://www.phe.gov/s3/dualuse/Documents/oversight-durc.pdf> [accessed 12.07.15].

[12] A framework for guiding US Department of Health and Human Services funding deci-sions about research proposals with the potential for generating highly pathogenic avian

influenza H5N1 viruses that are transmissible among mammals by respiratory droplets. US Department of Health and Human Services, 2013. Available from: <http://www.phe. gov/s3/dualuse/Documents/funding-hpai-h5n1.pdf> [accessed 12.07.15].

[13] Jackson RJ, Maguire DJ, Hinds LA, Ramshaw IA. Infertility in mice induced by a recombinant ectromelia virus expressing mouse zona pellucida glycoprotein 3. Biol Reprod 1998;58(1):152–9.

[14] Jackson RJ, Ramsay AJ, Christensen CD, Beaton S, Hall DF, Ramshaw IA. Expression of mouse interleukin-4 by a recombinant ectromelia virus suppresses cytolytic lymphocyte responses and overcomes genetic resistance to mousepox. J Virol 2001;75(3):1205–10.

[15] Herfst S, Schrauwen EJ, Linster M, et al. Airborne transmission of influenza A/H5N1 virus between ferrets. Science 2012;336(6088):1534–41.

[16] Imai M, Watanabe T, Hatta M, et al. Experimental adaptation of an influenza H5 HA confers respiratory droplet transmission to a reassortant H5 HA/H1N1 virus in ferrets. Nature 2012;486(7403):420–8.

[17] Severe acute respiratory syndrome. Wikipedia. Available from: <https://en.wikipedia.org/ wiki/Severe_acute_respiratory_syndrome> [accessed 18.07.15].

[18] Middle East respiratory syndrome. Wikipedia. Available from: <https://en.wikipedia.org/ wiki/Middle_East_respiratory_syndrome> [accessed 18.07.15].

[19] Cumulative number of confirmed human cases for avian influenza A(H5N1) reported to WHO, 2003-2015. World Health Organization, 2015. Available from: <http://www. who.int/influenza/human_animal_interface/EN_GIP_20150623cumulativeNumberH5N1 cases.pdf?ua=1%29> [accessed 18.07.15].

[20] Casadevall A, Relman DA. Microbial threat lists: obstacles in the quest for biosecurity? Nat Rev Microbiol 2010;8(2):149–54.

[21] U.S. Government gain-of-function deliberative process and research funding pause on selected gain-of-function research involving influenza, MERS, and SARS viruses. US Department of Health and Human Services, 2014. Available from: <http://www.phe.gov/ s3/dualuse/documents/gain-of-function.pdf> [accessed 18.07.15].

[22] Report of the Blue Ribbon Panel on influenza research. National Institute of Allergy and Infectious Diseases, 2007. Available from: <https://www.niaid.nih.gov/topics/flu/docu-ments/influenzablueribbonpanel2006.pdf>.

[23] Duprex WP, Fouchier RA, Imperiale MJ, et al. Gain-of-function experiments: time for a real debate. Nat Rev Microbiol 2015;13(1):58–64.

[24] Henkel RD, Miller T, Weyant RS. Monitoring select agent theft, loss and release reports in the United States—2004–2010. Appl Biosaf 2012;17(4):171–80.

[25] Pike RM. Laboratory-associated infections: summary and analysis of 3921 cases. Health Lab Sci 1976;13(2):105–14.

[26] Harding AL, Byers KB. Epidemiology of laboratory-associated infections. Fleming DO Fleming DO, Hunt DL, editors. Biological safety: principles and practices (3rd ed.). Washington, DC: ASM Press; 2000. p. 35–54.

[27] Respirator types. U.S. Department of Labor, Occupational Saftey and Health Administration. Available from: <https://www.osha.gov/video/respiratory_protection/ resptypes_transcript.html> [accessed 12.07.15].

[28] Klotz LC. Danger of potential-pandemic-pathogen research enterprises. MBio 2015;6(3):e00815.

[29] Fouchier RA. Studies on influenza virus transmission between ferrets: the public health risks revisited. MBio 2015;6:1.

[30] Shipp AC, Patterson AP. The National Institutes of Health system for enhancing the science, safety, and ethics of recombinant DNA research. Comp Med 2003;53(2):159–64.

[31] Young A, Penzenstadler N. Inside America's secretive biolabs. USA Today, 2015. Available from: <http://www.usatoday.com/longform/news/2015/05/28/biolabs-pathogens-location-incidents/26587505/> [accessed 16.07.15].

[32] Report on the Potential Exposure to Anthrax. Centers for Disease Control and Prevention, 2014. Available from: <http://www.cdc.gov/about/pdf/lab-safety/Final_Anthrax_Report.pdf> [accessed 16.07.15].

[33] Report on the inadvertent cross-contaimination and shipment of a laboratory specimen with influenza virus H5N1. Centers for Disease Control and Prevention, 2014. Available from: <http://www.cdc.gov/about/pdf/lab-safety/investigationcdch5n1contamination-eventaugust15.pdf> [accessed 16.07.15].

[34] Sisk, R. Pentagon Now Says Army Mistakenly Sent Live Anthrax to All 50 States. Miliatry.com, 2015. Available from: <http://www.military.com/daily-news/2015/09/01/pentagon-says-army-mistakenly-sent-live-anthrax-all-50-states.html>.

[35] Conclusion of select agent inquiry into *Burkholderia pseudomallei* release at Tulane National Primate Research Center. US Department of Health and Human Services, 2015. Available from: <http://www.cdc.gov/media/releases/2015/s0313-burkholderia-pseudomallei.html> [accessed 16.07.15].

[36] Palca, J. A CRISPR way to fix faulty genes. Shots, 2014. Available from: <http://www.npr.org/sections/health-shots/2014/06/26/325213397/a-crispr-way-to-fix-faulty-genes> [accessed 18.07.15].

[37] Pennisi E. The CRISPR craze. Science 2013;341(6148):833–6.

[38] Cyranoski D, Reardon S Chinese scientists genetically modify human embryos. Nature News, 2015. Available from: <http://www.nature.com/news/chinese-scientists-geneti-cally-modify-human-embryos-1.17378> [accessed 18.07.15].

[39] Tucker JB, Zilinskas RA. The promise and perils of synthetic biology. New Atlantis 2006;12:25–45.

[40] Moshelion M, Altman A. Current challenges and future perspectives of plant and agricul-tural biotechnology. Trends Biotechnol 2015;33(6):337–42.

[41] Jenkins C. Trends in United States biological materials oversight and institutional biosafety committees. J Res Adm 2014;45:11–46.

[42] Berger KM, Wolinetz C, McCarron K, You E, So KW, Hunt S Bridging science and security for Biological Research: Implementing the Revised Select Agents and Toxin Regulations Proceedings from the Meeting 22–23 April 2013. American Association for the Advancement of Science, Washington, DC. Available from: <https://www.aau.edu/WorkArea/DownloadAsset.aspx?id=14843> [accessed 10.07.15].

[43] Gottron F, Shea DA. Federal efforts to address the threat of bioterrorism: Selected issues and options for Congress. 2011. Congressional Research Service, Wasington, DC. Available from: <http://www.fas.org:8080/sgp/crs/terror/R41123.pdf> [accessed 10.07.15].

[44] National Research Council of the National Academies. Research universities and the future of America: Ten breakthrough actions vital to our nation's prosperity and security. National Academies Press, Wasinton, DC, 2012.

[45] Association of American Universities. Regulatory and financial reform of Federal Research Policy: Recommendations to the NRC Committee on Research Universities. Association of American Universities, Washington, DC. 2001. Available from: <http://eric.ed.gov/?id=ED517267> [accessed 15.06.15].

[46] Bosselman J, Myatt T. Doing more with less in biosafety programs. Environ Health Eng 2009;1(25). Available from: http://www.imakenews.com/bfoster/e_article001412009.cfm?x=bfsb140,b22Fh4yP.

[47] Foster RW. A creative solution for managing institutional biosafety committee research requests. Environ Health Eng 2007;1(19). Available from: http://www.imakenews.com/bfoster/e_article000975359.cfm?x=bbPvlGB%20,b22Fh4yP,w.

[48] Morse SA. Pathogen security-help or hindrance? Front Bioeng Biotechnol 2015;2:83.

[49] Shurtleff AC, Garza N, Lackemeyer M, Carrion Jr R, Griffiths A, Patterson J, et al. The impact of regulations, safety considerations and physical limitations on research progress at maximum biocontainment. Viruses 2012;4(12):3932–51.

[50] Sutton V. Smarter regulations: commentary on "Responsible conduct by life scientists in an age of terrorism". Sci Eng Ethics 2009;15(3):303–9.

[51] Schneider SL, Ness KK, Rockwell S, Shaver K, Brutkiewicz R. 2014. 2012 Faculty Workload Survey: Research Report. Federal Demonstration Partnership (FDP). Available from: <http://sites.nationalacademies.org/xpedio/groups/pgasite/documents/webpage/pga_087667.pdf> [accessed 06.09.2015].

[52] Sutton V. Survey finds biodefense researcher anxiety-over inadvertently violating regulations. Biosecur Bioterror 2009;7(2):225–6.

[53] Dias MB, Reyes-Gonzaleza L, Velosoa FM, Casman EA. Effects of the USA PATRIOT Act and the 2002 Bioterrorism Preparedness Act on select agent research in the United States. Proc Natl Acad Sci USA 2010;107(21):9556–61. Available from: http://www.pnas.org/content/107/21/9556.short.

[54] Gaudioso J, Salerno R. Science and government. Biosecurity and research: minimizing adverse impacts. Science 2004;304(5671):687.

[55] Race MS, Hammond E. An evaluation of the role and effectiveness of institutional biosafety committees in providing oversight and security at biocontainment laboratories. Biosecur Bioterror 2008;6(1):19–35.

[56] Association of American Medical Colleges. AAMC Report: Academic Medicine Investment in Medical Research. Association of American Medical Colleges, 2015. Available from: <http://www.huronconsultinggroup.com/Insights/Report/Education/AAMC_Investment_in_Medical_Research> [accessed 20.06.15].

[57] American Association for the Advancement of Science. Historical trends in Federal R&D. May 2015. Available from: <http://www.aaas.org/page/historical-trends-federal-rd> [accessed 06.07.15].

[58] National Institutes of Health. Research Portfolio Online Report Tools (RePORT). Available from: <http://report.nih.gov/index.aspx> [accessed 07.07.2015].

[59] National Center for Science and Engineering Statistics. Higher Education R&D Survey (HERD). Available from: <http://www.nsf.gov/statistics/srvyherd/#tabs-1>.

[60] Council on Government Relations Costing Committee. Finances of Research Universities. Available from: <http://www.cogr.edu/viewDoc.cfm?DocID=152058>; 2014 [accessed 09.06.15].

Index

Note: Page numbers followed by "*f*" and "*t*" refer to figures and tables, respectively.

Printed in the United States
By Bookmasters